TUMOR VASCULARIZATION

TUMOR VASCULARIZATION

Edited by

DOMENICO RIBATTI
Department of Basic Medical Sciences, Neurosciences and Sensory Organs, University of Bari Medical School, Bari, Italy

FRANCESCO PEZZELLA
Nuffield Division of Laboratory Science, Radcliffe Department of Medicine, University of Oxford, John Radcliffe Hospital, Oxford, United Kingdom

ELSEVIER

ACADEMIC PRESS
An imprint of Elsevier

Academic Press is an imprint of Elsevier
125 London Wall, London EC2Y 5AS, United Kingdom
525 B Street, Suite 1650, San Diego, CA 92101, United States
50 Hampshire Street, 5th Floor, Cambridge, MA 02139, United States
The Boulevard, Langford Lane, Kidlington, Oxford OX5 1GB, United Kingdom

British Library Cataloguing-in-Publication Data
A catalogue record for this book is available from the British Library

Library of Congress Cataloging-in-Publication Data
A catalog record for this book is available from the Library of Congress

ISBN: 978-0-12-819494-2

For Information on all Academic Press publications
visit our website at https://www.elsevier.com/books-and-journals

Publisher: Stacy Masucci
Acquisition Editor: Rafael Teixeira
Editorial Project Manager: Timothy Bennett
Production Project Manager: Punithavathy Govindaradjane
Cover Designer: Victoria Pearson

Typeset by MPS Limited, Chennai, India

Contents

List of contributors

Laetitia Andrique LAMC, Laboratoire de l'Angiogenèse et du Microenvironnement des Cancers (Inserm U1029), Pessac, France; Université de Bordeaux, Pessac, France

Andrea Barbieri Department of Molecular and Translational Medicine, University of Brescia, Brescia, Italy

Raymond Barnhill Department of Translational Research, Curie Institute, Paris, France

Andréas Bikfalvi LAMC, Laboratoire de l'Angiogenèse et du Microenvironnement des Cancers (Inserm U1029), Pessac, France; Université de Bordeaux, Pessac, France

Yun Cao Department of Pathology, Sun Yat-sen University Cancer Center, Guangzhou, P.R. China

Pedro García-Gómez Brain Metastasis Group, National Cancer Research Center (CNIO), Madrid, Spain

Jessica Guerra Department of Molecular and Translational Medicine, University of Brescia, Brescia, Italy

Adrian L. Harris Molecular Oncology Laboratories, Oxford University, Department of Oncology, Weatherall Institute of Molecular Medicine, John Radcliffe Hospital, Oxford, United Kingdom

Hynda Kleinman Department of Molecular Medicine and Biochemistry, The George Washington School of Medicine, Washington, DC, United States

Alia Komsany Nuffield Division of Laboratory Science, Radcliffe Department of Medicine, University of Oxford, John Radcliffe Hospital, Oxford, United Kingdom

Claire Lugassy Department of Translational Research, Curie Institute, Paris, France

Pierre Nassoy LP2N, Laboratoire Photonique Numérique et Nanosciences, Univ. Bordeaux, Talence, France; Institut d'Optique Graduate School & CNRS UMR 5298, Talence, France

Francesco Pezzella Nuffield Division of Laboratory Science, Radcliffe Department of Medicine, University of Oxford, John Radcliffe Hospital, Oxford, United Kingdom

Marco Presta Department of Molecular and Translational Medicine, University of Brescia, Brescia, Italy

Chao-Nan Qian Guangzhou Concord Cancer Center, Guangzhou, P.R. China

Gaelle Recher LP2N, Laboratoire Photonique Numérique et Nanosciences, Univ. Bordeaux, Talence, France; Institut d'Optique Graduate School & CNRS UMR 5298, Talence, France

Domenico Ribatti Department of Basic Medical Sciences, Neurosciences and Sensory Organs, University of Bari Medical School, Bari, Italy

Chiara Tobia Department of Molecular and Translational Medicine, University of Brescia, Brescia, Italy

Manuel Valiente Brain Metastasis Group, National Cancer Research Center (CNIO), Madrid, Spain

Peter Vermeulen Translational Cancer Research Unit, GZA Hospitals St Augustinus, University of Antwerp, Wilrijk-Antwerp, Belgium

Preface

The interaction between neoplastic cells and blood vessels, both newly formed or preexisting normal vessels, is one of the fundamental biological events involved in the development and progression of most solid and hematological tumors including the formation of metastases. Tumor angiogenesis is viewed as the consequence of an angiogenic switch, that is, a genetic event that endows the tumor with the ability to recruit blood vessels from the neighboring tissue. The newly formed tumor blood vessels have specific characteristics that allow discrimination from resting blood vessels. They are characterized by rapid proliferation, increased permeability, and disorganized architecture.

Initially thought to be a must for the grow and progression of tumors, the formation of new vessels was regarded as one of the hallmark of cancer. However, it was discovered that tumors can also growth without neoangiogenesis, mainly by co-opting preexisting vessels but also by vascular mimicry.

Since its discovery by Dr. Judah Folkman, tumor angiogenesis has been proposed as a target for novel tumor therapies. However, the success in the clinic of antiangiogenic compounds has been limited in contrast to many preclinical results obtained in animal models. This is in part due to the fact that tumors can be nonangiogenic, in part to several newly discovered mechanisms of resistance due both to the biology of the cancer cells and of the endothelium.

The field has therefore turned out to be more complex than previously thought.

We have attempted to create a book that will be of benefit not only to basic scientists working in this field, but also to clinicians by offering an overview of the field which now include the nonangiogenic tumors and the process vascular co-option.

We express our gratitude to all our colleagues who have contributed to this book covering some still controversial aspects of tumor angiogenesis.

Domenico Ribatti and Francesco Pezzella

CHAPTER

1

Sprouting and nonsprouting angiogenesis in tumors

Domenico Ribatti[1] and Francesco Pezzella[2]

[1]Department of Basic Medical Sciences, Neurosciences and Sensory Organs, University of Bari Medical School, Bari, Italy, [2]Nuffield Division of Laboratory Science, Radcliffe Department of Medicine, University of Oxford, John Radcliffe Hospital, Oxford, United Kingdom

Introduction

In 1971 Judah Folkman first advanced the hypothesis that tumor growth depends on the formation of new blood vessels from the preexisting vascular bed [1]. According to this hypothesis, endothelial cells must be switched from a resting state to a rapid growth phase by a diffusible chemical signal emanating from the tumor cells. New vessels in the body can be produced according to two main mechanisms: vasculogenesis and angiogenesis. Vasculogenesis occurs predominantly in the embryo and is characterized by the differentiation of hemoangioblasts into endothelial vessels. Vasculogenesis has been described in tumors and is discussed in Chapter 6. Angiogenesis is instead defined as a new blood vessel sprouting from a preexisting vessel, that is, capillaries and postcapillary venules [2].

Three types of classic angiogenic have been described: sprouting angiogenesis, the commonest observed in tumors, nonsprouting or intussusceptive microvascular growth (IMG) [3], and glomeruloid vascular proliferation [4]. Tumor growth starts with an avascular phase followed by a vascular phase (Fig. 1.1) [6]. The avascular phase appears to correspond to the histopathological picture presented by a small colony of neoplastic cells that reaches a steady state before it proliferates and becomes rapidly invasive. In this scenario, metabolites and catabolites are transferred by simple diffusion through the surrounding tissue.

Tumor Vascularization
DOI: https://doi.org/10.1016/B978-0-12-819494-2.00001-8

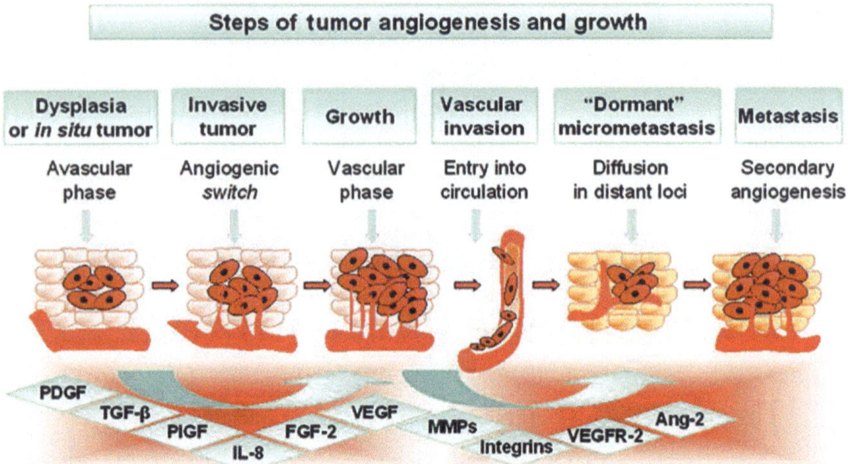

FIGURE 1.1 Steps of tumor angiogenesis and growth. Source: *Reproduced from Ribatti D, Vacca A. Overview of tumor angiogenesis. In: Figg WD, Folkman J, editors. Angiogenesis: an integrative approach from science to medicine. New York: Springer; 2008, p. 161–8 [5].*

Dormant tumors have been discovered during autopsies of individuals who died of causes other than cancer [7]. Carcinoma in situ is found in 98% of individuals aged 50–70 years who died of trauma, but is diagnosed in only 0.1% during life.

When tumors start to growth beyond the critical size of 2 mm at their site of origin they can ether undergo the angiogenic switch and move to induce new vessel growth or can keep growing remaining nonangiogenic by exploiting the host's preexisting vessels [8]. Sometime, as it has been show in the rat brain after an initial nonangiogenic phase, in which the tumor grows but only by exploiting preexisting vessels, a late angiogenic switch can occur [9].

In this chapter, we will discuss the angiogenic switch and the angiogenic processes while the nonangiogenic tumors will be described subsequently in Chapter 2.

The angiogenic switch

The mechanism of the switch was formally described by Hanahan, who developed transgenic mice in which the large "T" oncogene is hybridized to the insulin promoter [10]. In this model for β-islet cell tumorigenesis (RIP-Tag model), all islet cells in a transgenic mouse line express the large T antigen at birth. By 12 weeks, 75% of islets have progressed to small foci of proliferating cells, but only 4% are angiogenic and their number is closely correlated with the incidence of tumor formation [10]. The switch depends on increased production of one or more positive regulators of angiogenesis, such as vascular endothelial growth factor (VEGF), fibroblast growth factor-2 (FGF-2), interleukin-8 (IL-8), placental growth factor (PlGF), transforming growth factor-β (TGF-ß), platelet-derived growth factor (PDGF), pleiotrophins, and others, which cannot be held in check by the levels of antiangiogenic factors [6]. These can be exported from tumor cells, mobilized from the extracellullar matrix, or released from host cells recruited to the tumor. The switch clearly involves more than a simple upregulation of angiogenic activity and has thus been regarded as the result of the net balance between positive and negative regulators.

Sprouting tumor angiogenesis

In absence of pathological stimuli, vessels and their endothelial cells are quiescent and they remain so mostly because of a homeostatic mechanism relying on the Angiopoietin (Ang) family and their receptors. The first description of sprouting angiogenesis in tumor growth was reported by Ausprunk and Folkman [11]. They described the following stages: (1) the basement membrane is locally degraded on the side of the dilated peritumoral postcapillary venule situated closed to the angiogenic stimulus; (2) interendothelial contacts are weakened and endothelial cells migrate into the connective tissue; (3) a solid cord of endothelial cells form; and (4) lumen formation occurs proximal to the migrating front, contiguous tubular sprouts anastomose to form functionally capillary loops, parallel with the synthesis of the new basement membrane, and the recruitment of pericytes. In Ref. [12], Paku and Paweletz integrated the schema by Ausprunk and Folkman by means of ultrastructural observations, as follows: (1) a structural alteration of the basement membrane occurs, characterized by the loss of electron density over the entire circumference of the dilated mother vessel, followed by a partial degradation of the basement membrane at places where endothelial cell processes are projecting into the connective tissue; (2) endothelial cells migrate arranged in parallel, maintaining their basal-luminal polarity, forming a slit-like lumen and sealed by intact interendothelial junctions; (3) basement membrane is deposited continuously by the polarized endothelial cell while only the tip of the growing capillary bud is devoid of basement membrane; and (4) proliferating pericyte migrate along the basement membrane of the capillary bud, resulting in a complete coverage of the new vessel.

Initiation of sprouting require a growth signal usually an increased level of one of the VEGFs, as the basement membrane is degraded and

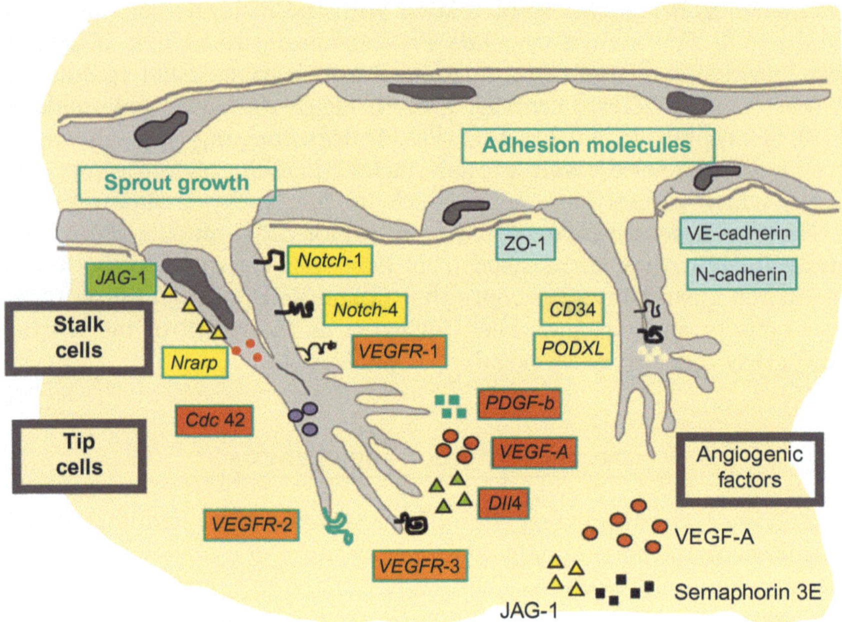

FIGURE 1.2 The functional specialization of endothelial cells during the sprouting process. VEGF-A induces the formation and extension of filopodia as well as the expression of Dll4 protein in the tip cells. Tip cells express high levels of Dll-4, PDGF-B, UNC5b, VEGFR-2, and VEGFR-3, and have low levels of Notch signaling activity. During both mouse and zebrafish angiogenesis, VEGFR-3 is most strongly expressed in the leading tip cell and is downregulated by Notch signaling in the stalk cell. Stalk cells produce fewer filopodia, are more proliferative, form tubes and branches, and form a vascular lumen. They also establish junctions with neighboring cells and synthesize basement membrane components. Stalk cells have high levels of Notch signaling activity and elevated expression of JAG-1. Nrarp is a downstream of Notch that counteracts Notch signaling and is expressed in stalk cells at branch points. Adherens junction formation is associated with the inhibition of endothelial cell migration in monolayers. This process is mediated by VE-cadherin. Another cell adhesion molecule expressed in endothelial cells is the N-cadherin. VE-cadherin is strictly required for the polarization of endothelial cells. Endothelial cell polarization starts with the delivery of deadhesive apical glycoproteins to the cell–cell contact via exocytosis. Deadhesive molecules include CD34-sialomucins, such as CD34 and PODXL. Source: *Reproduced from Ribatti D, Crivellato E. "Sprouting angiogenesis": a reappraisal. Dev Biol 2012;372:157–65 [13].*

the endothelial cells start to detach, the vessels becomes leaky. As the endothelial cells proliferate, they migrate in the surrounding stroma and they arrange themselves into a stalk. As the length of the cell row increases, the newly formed endothelium differentiates into tip and stalk cells bearing different morphologies and functional properties (Fig. 1.2). Endothelial tip cells primarily migrate but proliferate only minimally, in contrast to the proliferating endothelial stalk cells [14].

Tip cells have been compared to axonal growth cones during neurite outgrowth. The phenotypic specialization of endothelial cells as tip or stalk cells is very transient and reversible, depending on the balance between proangiogenic factors, such as VEGF and Jagged-1 (JAG-1), and suppressors of endothelial cell proliferation, such as delta-like ligand 4 (Dll-4)-Notch activity [15,16]. Endothelial cells are exposed to VEGF-A which leads to an activation of VEGF receptor-2 (VEGFR-2) and Dll4 in exposed cells. Lateral inhibition between activated cells mediated by Delta-Notch signaling enhances relative differences between neighboring cells and triggers the formation of a single tip cell [13]. The stalk cells induce deposit of basement membrane and recruit pericytes while becoming quiescent, thus stabilizing the newly formed vessel.

During the transition from active sprouting to quiescence endothelial cells, tip cell adopts a "phalanx" phenotype (so-called as the endothelial cells form an ordered monolayer reminiscent of the military "phalanx formation" of the ancient Greek soldiers) that acquires a lumen, nonproliferating, and immobile cells, which promotes vessel integrity and stabilizes the vasculature through increased cell adhesion and dampened response to VEGF (Fig. 1.2) [17].

Lumen formation, essential to allow a flow of blood through the new vessel, involves a complex molecular mechanism composed of endothelial cell repulsion at the cell–cell contacts within the endothelial cell cords, junctional rearrangement, and endothelial cell shape change [18]. After the vascular lumen has been established, blood initiates to flow through the newly formed vessel. As angiogenesis progress, some of newly formed vessels may become redundant and regresses: this process is called remodeling.

Remodeling involves the regression of some of the newly formed vessels as well as changes in the diameter of vessel lumens and vascular wall thickening. Remodeling determines the formation of large and small vessels, the establishment of directional flow, the association with mural cells (pericytes and smooth muscle cells), and the adjustment of vascular density to meet the nutritional requirements of the surrounding tissue. Pruning involves the removal of excess of endothelial cells which form redundant channels and occurs physiologically during embryonic development and during corpus luteum regression, pathologically in intratumor angiogenesis and upon therapeutic VEGF manipulation [19]. In fact, in both preclinical and clinical settings, anti-VEGF drugs induce remodeling of tumor blood vessels leading to a more normalized vasculature [20]. As a consequence of the structural and functional normalization of tumor blood vessels, blood flow is increased and cytotoxic drugs can more easily be delivered to the tumor.

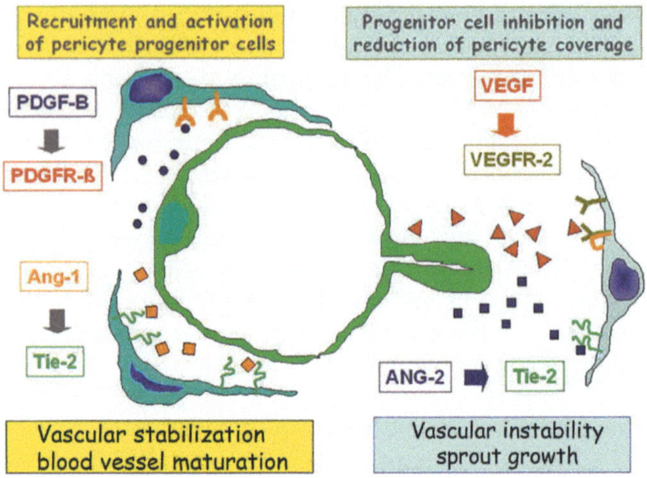

FIGURE 1.3 Schematic drawing that illustrates the paracrine interactions occurring between pericyte precursor cells and endothelial cells in PDGF-mediated angiogenesis. Endothelial cells secrete PDGF-B that causes pericyte precursor cell proliferation and migration through activation of PDGFR-β. Pericytes surround and cover early endothelial tubes. By contrast, endothelial cells in vascular sprouts release VEGF, which in turn mediates suppression of PDGFR-β signaling through the induction of VEGFR-2/PDGFR-β complexes. This pathway abrogates pericyte coverage of endothelial sprouts leading to vascular instability and regression. Source: *Reproduced from Ribatti D, Nico B, Crivellato, E. The role of pericytes in angiogenesis. Int J Dev Biol 2011;55:261–8.*

Pruning and remodeling of the vascular network may be stimulated by tissue-derived signaling molecules and blood flow conditions (e.g., wall-shear stress and pressure). The stabilization of the newly formed vessel and the maintenance of the existing vasculature are late events in the angiogenic process. Pericyte adhesion to native capillaries and endothelial cell wrapping by surrounding pericytes are basic events in blood vessel stabilization and maturation (Fig. 1.3) [21]. Pericytes provide structural support for the capillary wall, acting as a scaffold along which endothelial cells migrate during sprouting. Reduced pericyte coverage is associated with metastasis, and overexpression of PDGF-B increases pericyte coverage and inhibits tumor growth.

Intussusceptive microvascular growth angiogenesis

IMG is an alternative or additional mechanism of capillary growth whereby the vascular network expands by insertion of newly formed columns of interstitial tissue structures into the vascular lumen called tissue pillars or posts. The first reports on IMG were published by Burri

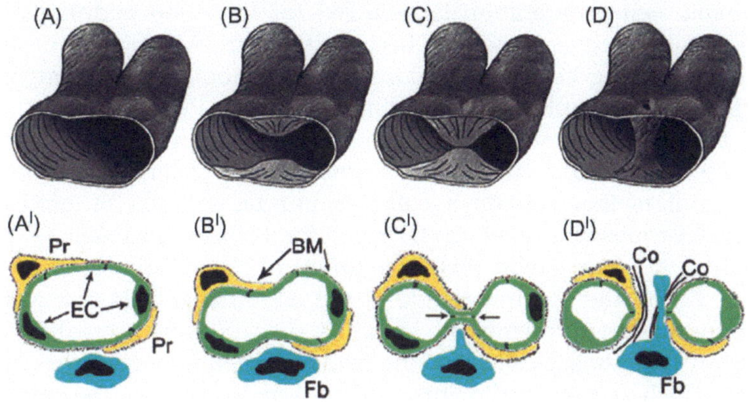

FIGURE 1.4 3D (A–D) and 2D (A′–D′) scheme depicting the generation of transluminal pillar by intussusceptive angiogenesis. Simultaneously protrusion of opposing capillary walls into the vessel lumen (A, B; A′, B′) results in creation of interendothelial contact zone (C; C′). In a subsequent step, the endothelial bilayer becomes perforated and the newly formed pillar core got invaded by fibroblasts (Fb) and pericytes (Pr), which lay down collagen fibrils (Co in D′). Source: *Reproduced from Ribatti D, Djonov V. Intusussceptive microvascular growth in tumors. Cancer Lett 2012;316:126–31.*

and coworkers who investigated the lung vasculature in postnatal rats using corrosion casting and scanning electron microscopy [22,23]. Transluminal pillar formation is one particular way of expanding vessels [24]. According to Burri's group, this mechanism proceeds through four steps: (1) protrusion of opposing capillary walls into the lumen and the creation of a contact zone between facing endothelial cells; (2) reorganization of their intercellular junctions and central perforation of the endothelial bilayer; (3) formation of an interstitial pillar core by invading supporting cells (myofibroblasts, pericytes) and deposition of matrix, such pillars ranging in diameter from 1 to 2.5 μm; and (4) enlargement in thickness of the pillars without additional qualitative alteration (Fig. 1.4) [25].

The molecular mechanism of IMG is not fully understood, but shear stress plays a major role, as well as increasing blood flow rate. The role of hemodynamic in control of IMG has been demonstrated by clamping of one of the dichotomous branches of an artery in the developing chorioallantoic membrane of the chick embryo [26]. Increase in blood flow and pressure in the cognate artery resulted in an almost immediate effect on branching morphology with pillars beginning to appear in 15–30 minutes of clamping [26].

IMG generates vessels more rapidly with a less metabolic demand as compared to sprouting angiogenesis and is a putative strategy that tumors can use for rapid adaptation to milieu changes. In fact, no endothelial cell proliferation is required for IMG and endothelial cells only increase their volume and become thinner. IMG occurs in several tumors, including colon and mammary carcinomas, melanoma, B-cell non-Hodgkin's lymphoma, and glioma [27–31]. Paku et al. [32] elucidated the mechanism of pillar formation in experimental tumors. By using electron and confocal microscopy, they observed intraluminal nascent pillars that contain a collagen bundle covered by endothelial cells and proposed a new mechanism for the development of pillars consisting of four steps: (1) formation of intraluminal endothelial bridges; (2) on the abluminal side of the endothelial cells that form the bridge the basement membrane is locally disrupted by proteolytic activity; (3) an endothelial cell from the bridge adheres to a nearly collagen bundle which is transferred through the lumen, reaches the other side of the lumen, and is transferred into the connective tissue on the other side of the vessel; and (4) further pillar maturation occurs through the immigration of fibroblasts/myofibroblasts and pericytes into the pillar and subsequent extracellular matrix proteins (collagen and fibrin) deposition by these cells.

A switch from sprouting to intussusceptive angiogenesis might represent an adaptive response to treatment with various antitumor and antiangiogenic compounds to restore the hemodynamic and structural properties of the vasculature enhancing tumor drug delivery and sensitivity to treatments.

In a hepatocellular carcinoma experimental model treated with sirolimus, a mTOR inhibitor, during the treatment and the early recovery phase vascular sprouting, was absent, whereas IMG was observed [33]. Whereas the capillary plexus expanded primarily by sprouting in the control animals, this mechanism was nearly absent in the tumors of treated animals and replaced by intussusception [33]. Radiotherapy of mammary carcinoma allografts or treatment with an inhibitor of VEGF tyrosine kinase (PTK787/ZK222854) results in transient reduction in tumor growth rate with decreased tumor vascularization followed by posttherapy relapse with extensive IMG, characterized by a plexus composed of enlarged sinusoidal-like vessels containing multiple transluminal tissue pillars, a decrease in the intratumoral microvascular density, probably as a result of intussusceptive pruning associated to a minimal reduction of the total microvascular exchange area [34]. Moreover, the switch to IMG improves the perfusion of the tumor mass as has been shown by an improvement in oxygen supply of the tumor mass [34]. Tumor recovery after treatment with inhibitors of VEGFR signaling in the RIP-Tag 2 and Lewis lung carcinoma models was associated with rapid revascularization and evidence of IMG [35]. Similar changes in vasculature have been observed in a murine renal cell carcinoma study [36], in murine orthotopic B16/BL6 melanoma tumor model after treatment with either PTK/ZK [37] or other tyrosine kinase inhibitors [38–40]. Restoration of glomerular capillary structure after induction of Thy1.1 nephritis occurred by intussusceptive angiogenesis and this in spite of the VEGFR-2 and PDGFRβ inhibition [41]. The results provide an important insight into IMG control mechanisms and indicate differential molecular regulation between sprouting and intussusceptive angiogenesis.

Glomeruloid vascular proliferation

Small vascular glomeruloid bodies, so-called because of their morphologic resemblance of the renal glomeruli, first described in glioblastomas, have been seen, although less frequently, in other tumors as well [42]. They are formed by groove of small vessels with a basement membrane and irregular pericyte coverage [43,44]. Dome et al. [42] raised the hypothesis that there could be two types of glomeruloid bodies. The first, formed by an "active" mechanism would be the one in which angiogenesis actually occurs and the glomeruloid vessels are newly formed, possibly because of the action of VEGF-A [4]. The second type or "passive" is one in which no new vessels are formed but preexisting capillaries are coiled and folded by metastatic cells which extravasate and then adhere to the abluminal surface of the capillaries and pulling them into a glomeruloid shape [42].

Concluding remarks

Unlike physiological angiogenesis, tumor blood vessels are structurally abnormal, are characterized by reduced blood flow, leakiness, and dilatation, and lack pericyte coverage. The hierarchy of arterioles, capillaries, and venules associate with normal vasculature is absent in that of tumors. The hyperpermeable tumor vasculature allows tumor intravasation and systemic dissemination [20].

These changes lead to a hypoxic condition and an accumulation of metabolic wastes [45]. Hypoxia has important effects on tumor response to radiation and chemotherapy, generally reducing their effectiveness. Hypoxia promotes the recruitment of bone marrow–derived cells (BMDCs), tumor-associated macrophages (TAMs), Tie-2-expressing monocytes (TEMs), neutrophils, and mast cells, CD11b-positive GR-1-positive cells, all able to release proangiogenic factors [46]. In turn, cancer stem cells produce higher levels of VEGF in hypoxic conditions [47] and can recruit other endothelial cell precursors able to further stimulate angiogenesis [48]. The use of hypoxia inducible factor-1 alpha

(HIF-1α) inhibitors to block tumor angiogenesis has therefore received attention. For example, topotecan in combination with bevacizumab reduces HIF-1α promotes tumor regression in preclinical models [49].

VEGF blockade inhibits sprouting angiogenesis, but is not efficient in suppressing other modes of tumor vascularization, involving the recruitment of BMDCs [50], vessel co-option, or vasculogenic mimicry. Moreover, after treatment with VEGF blockers, tumor cells survive in poorly oxygenated niches and adapt to antiangiogenesis by increasing different cellular survival processes [51].

Antiangiogenic drugs are more efficient in well-vascularized cancers as clear cell renal cancer [52], whereas they are less effective in less vascularized cancers as pancreatic adenocarcinoma and gastric cancer [53,54].

As Jain said: "We need to develop therapeutic agents that normalize the entire tumor microenvironment, including immune and other stromal cells, and not just the tumor blood vessels" [55].

References

[1] Folkman J. Tumour angiogenesis: therapeutic implications. N Engl J Med 1971;285:1182–6.
[2] Risau W. Mechanisms of angiogenesis. Nature 1997;386:671–4.
[3] Ribatti D. Genetic and epigenetic mechanisms in the early development of the vascular system. J Anat 2006;208:139–52.
[4] Sundberg S, Nagy JA, Brown LF, Feng D, Eckelhoefer IA, Manseau EJ, et al. Glomeruloid microvascular proliferation follows adenoviral vascular permeability factor/vascular endothelial growth factor-164 gene delivery. Am J Pathol 2001;158:1145–60.
[5] Ribatti D, Vacca A. Overview of tumor angiogenesis. In: Figg WD, Folkman J, editors. Angiogenesis: an integrative approach from science to medicine. New York: Springer; 2008. p. 161–8.
[6] Ribatti D, Nico B, Crivellato E, Roccaro AM, Vacca A. The history of the angiogenic switch concept. Leukemia 2007;21:44–52.
[7] Black WC, Welch HG. Advances in diagnostic imaging and overestimations of disease prevalence and the benefits of therapy. N Engl J Med 1993;328:1237–43.
[8] Pezzella F, Pastorino U, Tagliabue E, Andreola S, Sozzi G, Gsaparini G, et al. Non-small lung carcinoma tumor growth without morphological evidence of neo-angiogenesis. Am J Pathol 1997;151:1417–23.
[9] Holash J, Maisonpierre PC, Compton D, Boland P, Alexander CR, Zagzag D, et al. Vessel cooption, regression and growth in tumors mediated by angiopoietins and VEGF. Science 1999;284:1994–8.
[10] Hanahan D. Heritable formation of pancreatic beta-cell tumors in transgenic mice expressing recombinant unsylin/simian virus 40 oncogene. Nature 1985;315:115–22.
[11] Ausprunk DH, Folkman J. Migration and proliferation of endothelial cells in preformed and newly formed blood vessels during tumor angiogenesis. Microvasc Res 1977;14:53–65.
[12] Paku S, Pawoletz N. First steps of tumor-related angiogenesis. Lab Invest 1991;65:334–46.

[13] Ribatti D, Crivellato E. "Sprouting angiogenesis": a reappraisal. Dev Biol 2012;372:157–65.
[14] Gerhardt H, Golding M, Fruttiger M, Ruhrberg C, Lundkvist A, Abramsson A, et al. VEGF guides angiogenic sprouting utilizing endothelial tip cell filopodia. J Cell Biol 2003;161:1163–77.
[15] Eilken HM, Adams RH. Dynamics of endothelial cell behavior in sprouting angiogenesis. Curr Opin Cell Biol 2010;22:617–25.
[16] Geudens I, Gerhardt H. Coordinating cell behavior during blood vessel formation. Development 2011;138:4569–83.
[17] Bautch VL. Endothelial cells form a phalanx to block metastasis. Cell 2009;136:810–12.
[18] Iruela-Arispe ML, Davis GE. Cellular and molecular mechanisms of vascular lumen formation. Dev Cell 2009;16:222–31.
[19] Inai T, Mancuos M, Hashizume H, Baffert F, Haskell A, Baluk P, et al. Inhibition of vascular endothelial growth factor (VEGF) signaling in cancer causes loss of endothelial fenestrations, regression of tumor vessels, and appearance of basement membrane ghosts. Am J Pathol 2004;165:35–52.
[20] Jain RK. Normalization of tumor vasculature. An emerging concept in antiangiogenic therapy. Science 2005;307:58–62.
[21] Ribatti D, Nico B, Crivellato E. The role of pericytes in angiogenesis. Int J Dev Biol 2011;55:261–8.
[22] Burri PH, Tarek MR. A novel mechanism of capillary growth in the rat pulmonary microcirculation. Anat Rec 1990;228:35–45.
[23] Caduff JH, Fischer LC, Burri PH. Scanning electron microscopic study of the developing microvasculature in the postnatal rat lung. Anat Rec 1986;216: 154–64.
[24] Djonov V, Schnid M, Tschanz SA, Burri PH. Intussusceptive angiogenesis: its role in embryonic vascular network formation. Circ Res 2000;86:286–92.
[25] Ribatti D, Djonov V. Intusussceptive microvascular growth in tumors. Cancer Lett 2012;316:126–31.
[26] Djonov V, Kurz H, Burri PH. Optimality in the developing vascular system: branching remodeling by means of intussusception as an efficient adaptation mechanism. Dev Dyn 2002;224:391–402.
[27] Djonov V, Hogger K, Sedlacek R, Laissue J, Draeger A. MMP-19: cellular localization of a novel metalloproteinase within normal breast tissue and mammary gland tumours. J Pathol 2001;195:147–55.
[28] Nico B, Crivellato E, Guidolin D, Annese T, Longo V, Finato N, et al. Intussusceptive microvascular growth in human glioma. Clin Exp Med 2010;10:93–8.
[29] Patan S, Munn LL, Jain RK. Intussusceptive microvascular growth in a human colon adenocarcinoma xenograft: a novel mechanism of tumor angiogenesis. Microvasc Res 1996;51:260–72.
[30] Ribatti D, Nico B, Floris C, Mangieri D, Piras F, Ennas MG, et al. Microvascular density, vascular endothelial growth factor immunoreactivity in tumor cells, vessel diameter and intussusceptive microvascular growth in primary melanoma. Oncol Rep 2005;14:81–4.
[31] Crivellato E, Nico B, Vacca A, Ribatti D. B-cell non-Hodgkin's lymphomas express heterogeneous patterns of neovascularization. Haematologica 2003;88: 671–8.
[32] Paku S, Deszo K, Bugyik E, Tovari J, Timar J, Nagy P, et al. A new mechanism for pillar formation during tumor-induced intussusceptive angiogenesis: inverse sprouting. Am J Pathol 2011;179:1573–85.

[33] Semela D, Piguet AC, Kolev M, Schmitter K, Hlushechuk R, Djonov V, et al. Vascular remodeling and antitumoral effects of mTOR inhibition in a rat model of hepatocellular carcinoma. J Hepatol 2007;46:840–8.
[34] Hlushchuk R, Riesterer O, Baum O, Wood J, Gruber G, Pruschy M, et al. Tumor recovery by angiogenic switch from sprouting to intussusceptive angiogenesis after treatment with PTK787/ZK222584 or ionizing radiation. Am J Pathol 2008;173:1173–85.
[35] Mancuso MR, Davis R, Norberg SM, O'Brien S, Sennino B, Nakahara T, et al. Rapid vascular regrowth in tumors after reversal of VEGF inhibition. J Clin Invest 2006;116:2610–21.
[36] Drevs J, Muller-Driver R, Wittig C, Fuxius S, Esser N, Hugenschmidt H, et al. PTK787/ZK 222584, a specific vascular endothelial growth factor-receptor tyrosine kinase inhibitor, affects the anatomy of the tumor vascular bed and the functional vascular properties as detected by dynamic enhanced magnetic imaging. Cancer Res 2002;62:4015–22.
[37] Rudin M, McSheehy PM, Allegrini PR, Rausch M, Baumann D, Becquet MK, et al. PTK787/ZK222584, a tyrosine kinase inhibitor of vascular endothelial growth factor receptor, reduces uptake of the contrast agent GdDOTA by murine orthotopic B16/BL6 melanoma tumours and inhibits their growth in vivo. NMR Biomed 2005;18:308–21.
[38] Nakamura K, Yamamoto A, Kamishohara M, Takahashi K, Taguchi E, Miura M, et al. KRN633: a selective inhibitor of vascular endothelial growth factor receptor-2 tyrosine kinase that suppresses tumor angiogenesis and growth. Mol Cancer Ther 2004;3:1639–49.
[39] Nakamura K, Taguchi E, Miura T, Yamamoto A, Takahashi K, Bichat F, et al. KRN951, a highly potent inhibitor of vascular endothelial growth factor receptor tyrosine kinases, has antitumor activities and affects vascular functional properties. Cancer Res 2006;66:9134–42.
[40] Ruggeri B, Singh J, Gingrich D, Angeles T, Albom M, Yang S, et al. CEP-7055: a novel, orally active pan inhibitor of vascular endothelial growth factor receptor tyrosine kinases with potent antiangiogenic activity and antitumor efficacy in preclinical models. Cancer Res 2003;63:5978–91.
[41] Wnuk M, Hlushchuk R, Tuffin G, Huynh-Do U, Djonov V. The effects of PTK787/ZK222584, an inhibitor of VEGFR and PDGFRβ pathways, on intussusceptive angiogenesis and glomerular recovery from Thy1.1 nephritis. Am J Pathol 2011;178:1899–912.
[42] Dome B, Timar J, Paku S. A novel concept of glomeruloid body formation in experimental cerebral metastases. J Neuropathol Exp Pathol 2003;62:655–61.
[43] Rojiani AM, Dorovini-Zis K. Glomeruloid vascular structures in glioblastoma multiforme: an immunohistochemical and ultrastructural study. J Neurosurg 1996;85:1078–84.
[44] Wesseling P, Vandersteenhoven JJ, Downey BT, Ruiter DJ, Burger PC. Cellular components of microvascular proliferation in human glial and metastatic brain neoplasms. A light microscopic and immunohistochemical study of formalin-fixed, routinely processed material. Acta Neuropathol 1993;85:508–14.
[45] De Bock K, Cauwenberghs S, Carmeliet P. Vessel abnormalization: another hallmark of cancer? Molecular mechanisms and therapeutic implications. Curr Opin Genet Dev 2011;21:73–9.
[46] Ferrara N. Pathways mediating VEGF-independent tumor angiogenesis. Cytokine Growth Factors Rev 2010;21:21–6.
[47] Bao S, Wu Q, Sathornsumetee S, Hao Y, Li Z, Hjelmeland AB, et al. Stem cell-like glioma cells promote tumor angiogenesis through vascular endothelial growth factor. Cancer Res 2006;66:7843–8.

[48] Folkins C, Shaked Y, Man S, Tang T, Lee CR, Zhu Z, et al. Glioma tumor stem-like cells promote tumor angiogenesis and vasculogenesis via vascular endothelial growth factor and stromal-derived factor 1. Cancer Res 2009;69:7243–51.

[49] Rapisarda A, Hollingshead M, Uranchimeg B, Bonomi CA, Borgel SD, Carter JP, et al. Increased antitumor activity of bevacizumab in combination with hypoxia inducible factor-1 inhibition. Mol Cancer Ther 2009;8:1867–77.

[50] Shaked Y, Ciarrocchi A, Franco M, Lee CR, Man S, Cheung AM, et al. Therapy-induced acute recruitment of circulating endothelial progenitor cells to tumors. Science 2006;313:1785–7.

[51] Rohwer N, Cramer T. Hypoxia-mediated drug resistance: novel insights on the functional interaction of HIFs and cell death pathways. Drug Resist Update 2011;14:191–201.

[52] Yang JC, Haworth L, Sherry RM, Hwu P, Schwartzentruber DJ, Topalian SL, et al. A randomized trial of bevacizumab, an anti-vascular endothelial growth factor antibody, for metastatic renal cancer. N Engl J Med 2003;349:427–34.

[53] Kindler HL. Pancreatic cancer: an update. Curr Oncol Rep 2007;9:170–6.

[54] Ohtsu A, Shah MA, Van Cutsem E, Rha SY, Sawaki A, Park SR, et al. Bevacizumab in combination with chemotherapy as first-line therapy in advancer gastric cancer: a randomized, double-blind, placebo-controlled phase III study. J Clin Oncol 2011;29:3968–76.

[55] Jain RK. Antiangiogenesis strategies revisited: from starving tumors to alleviating hypoxia. Cancer Cell 2014;26:605–22.

CHAPTER

2

Nonangiogenic tumor growth

Peter Vermeulen[1] *and Francesco Pezzella*[2]

[1]Translational Cancer Research Unit, GZA Hospitals St Augustinus, University of Antwerp, Wilrijk-Antwerp, Belgium, [2]Nuffield Division of Laboratory Science, Radcliffe Department of Medicine, University of Oxford, John Radcliffe Hospital, Oxford, United Kingdom

Introduction

Tumors can grow without inducing angiogenesis [1]. The main mechanism by which neoplasms obtain a vasculature in the absence of angiogenesis is by exploiting, or co-opting, normal preexisting vessels [2]. A less common mechanism in which the cancer cells themselves form blood-perfused channels is called vasculogenic mimicry (discussed in Chapter 5). We focus in this chapter on vessel co-option, sometimes also called vascular co-option, angiotropism, or pericytic mimicry.

Identification of nonangiogenic tumors

Nonangiogenic tumors have historically been identified by histology. The first description was done in lung tumors. Four types of nonangiogenic growth have been observed in the lung: alveolar, lepidic, interstitial, and perivascular.

Alveolar growth pattern

Lung tissue is characterized by the presence of small air spaces, the alveolar spaces, defined by the alveolar walls. The alveolar walls have capillaries supported by fibroblasts. The alveolar spaces are lined by epithelial cells, type 1 (the most common type) and type 2 pneumocytes (Fig. 2.1A).

Tumor Vascularization
DOI: https://doi.org/10.1016/B978-0-12-819494-2.00002-X

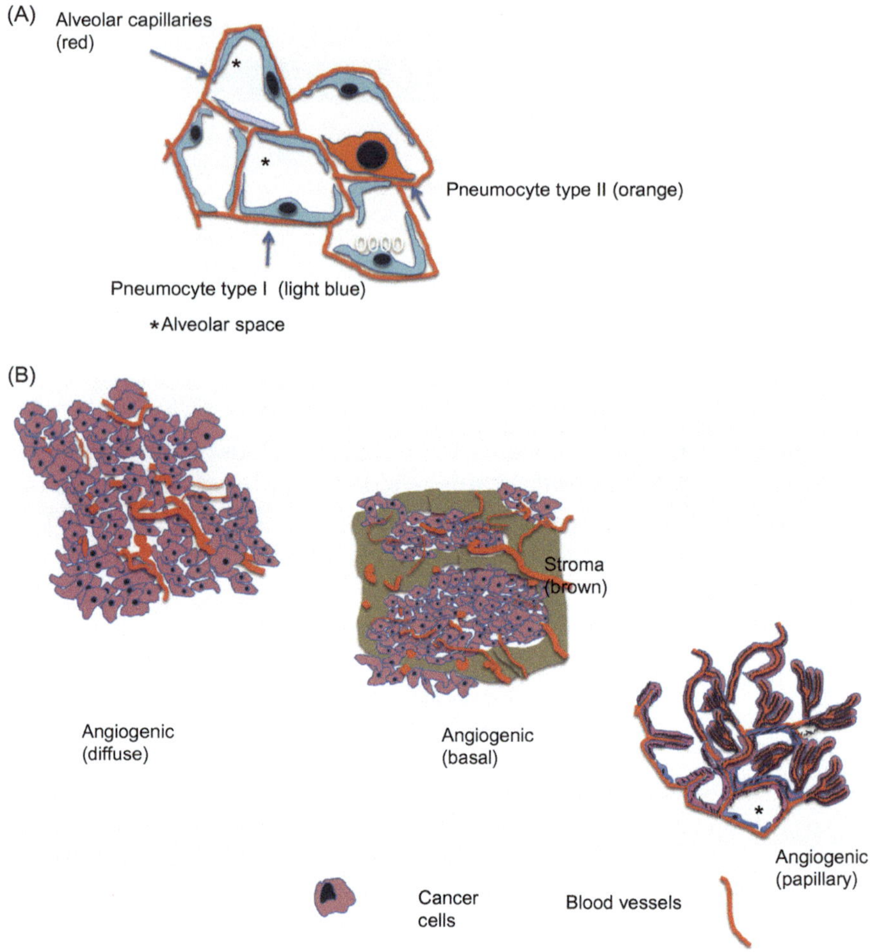

FIGURE 2.1 Vascular patterns in lung tumors. (A) Structure of the normal lung parenchyma. The alveolar spaces, containing air, are delimited by alveolar walls. These are lined on the inside by pneumocytes of type 1 and type 2. In the alveolar walls, lie the capillaries in which blood flows releasing CO_2 and taking up oxygen. (B) Three angiogenic patterns can be seen both in primary and metastatic tumors. In the first pattern, or diffuse growth pattern, the normal tissue is replaced by a chaotic mixture of cancer cells, vessels, and variable amount of stroma, without any specific architecture. In the second pattern, the basal pattern, collections of variable size of cancer cells are surrounded by fibrotic desmoplastic bands containing newly formed vessels which can also extend among the cancer cells. In the papillary pattern, corresponding to the lepidic angiogenic pattern of the WHO classification, numerous papillae formed by a thin fibrovascular stalk covered by a layer of neoplastic cells, sprout from the alveolar capillaries, and remodel the preexisting architecture. (C) Nonangiogenic, vessel co-opting patterns. In the alveolar pattern, the normal lung parenchyma is preserved and the alveolar spaces are filled by cancer cells, and in a later stage the pneumocytes disappear. In the alveolar lepidic pattern, corresponding to the lepidic nonangiogenic pattern of the WHO classification, the cancer cells grow on the alveolar wall as a single layer replacing the pneumocytes. In the nonangiogenic interstitial pattern, observed mainly in metastatic lesions, the cancer cells extravasate but, instead of invading the air space, grow in the stroma that surrounds the co-opted capillaries, in the space between the vessel and the overlying pneumocytes. Finally, in the perivascular cuffing pattern, the malignant cells grow along larger vessels like arteriolae.

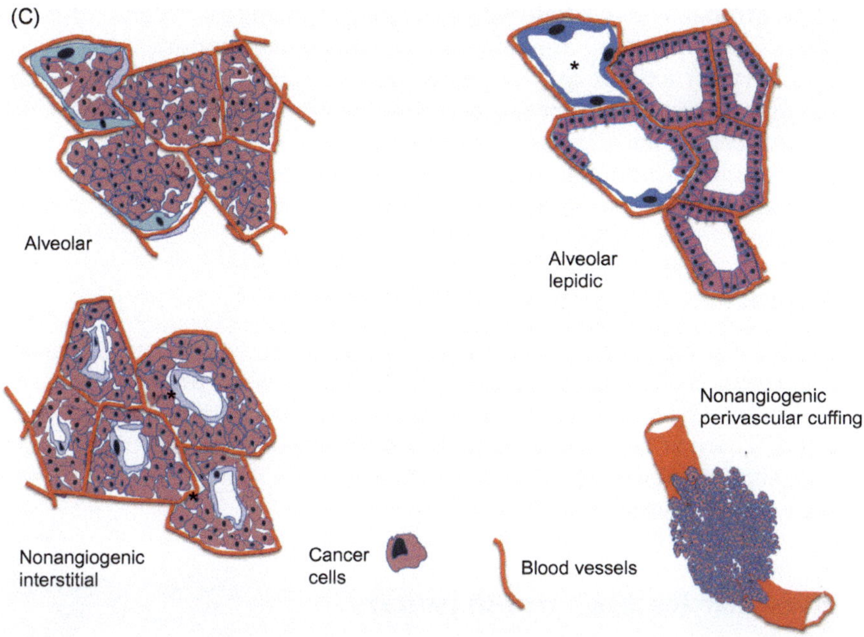

FIGURE 2.1 *(Continued)*

As in any other organ, the angiogenic tumors in the lung are characterized by destruction of the tissue architecture which is replaced by a neoplastic mass containing newly formed vessels arranged in a chaotic pattern (Fig. 2.1B).

In nonangiogenic tumors, on the contrary (Fig. 2.1C), the underlying structure of the host organ, in particular the vasculature, is preserved. In case of the lung, the "chicken wire" appearance of the normal vessel arrangement is maintained. This is because both primary and secondary nonangiogenic lung tumors can grow by filling the alveolar spaces [3,4]. Initially the pneumocytes which line the alveolar walls are still present but eventually they disappear as the neoplastic cells infiltrate between the pneumocytes and the basal membrane causing the pneumocytes to detach and subsequently die [5,6]. The neoplastic cells colonize nearby tissue by migrating from one alveolar space to another through the alveolar pores [7].

Lepidic growth pattern

The lepidic growth pattern differs from the alveolar growth pattern in that the neoplastic cells do not fill the alveolar spaces but instead line the alveolar wall by replacing the pneumocytes. The cancer cells are usually arranged in a single layer covering the alveolar walls. There is no associated sprouting angiogenesis in this growth pattern.

Papillary structures can however protrude into the alveolar lumen. These structures are composed of a newly formed blood vessel covered by cancer cells. This angiogenic growth pattern is called "papillary."

The lepidic tumors used to belong to the bronchioloalveolar carcinomas group, a now abandoned nomenclature [8].

Interstitial growth pattern

In this growth pattern, the cancer cells occupy the interstitial stroma that surrounds the blood vessels of the alveolar walls and do not enter the alveolar spaces. Instead, these spaces are left intact and are compressed. There is no angiogenesis and the chicken wire vascular pattern of the lung is preserved. This growth pattern is largely limited to secondary lung tumors.

Perivascular (cuffing) growth pattern

Rather than co-opting the small capillaries of the alveolar walls in the alveolar and lepidic growth pattern, cancer cells can also grow as a multilayered cuff around larger veins or arteries of the lung. This fourth nonangiogenic growth pattern is also mainly observed in metastatic tumors in the lung [6].

Vessel co-option in the lung can be recognized by the following histomorphological characteristics:

- The architecture of the lung vasculature is preserved, as discussed above. This is also reflected by the presence of anthracotic pigment in the connective tissue that surrounds the co-opted blood vessels and by the expression of the LH39 antigen in the basal membrane of these mature vessels—LH39 is absent in newly formed blood vessels [9].
- Weak and patchy expression of $\alpha_v\beta_3$ integrins, similar to the expression pattern of the endothelial cells of the normal lung [10].
- Within the alveoli that are filled with cancer cells, there are no endothelial cells nor is there any desmoplastic reaction.
- Comparable to the normal lung, there is only a very small fraction of proliferating endothelial cells [11].
- Tumors having both angiogenic and nonangiogenic areas are commonly seen [4].

In the liver, careful histopathological examination of colorectal and breast cancer metastases has shown that these tumors can present with three main histopathological growth patterns (HGP) [12,13] (Fig. 2.2). The major discriminating feature between these growth

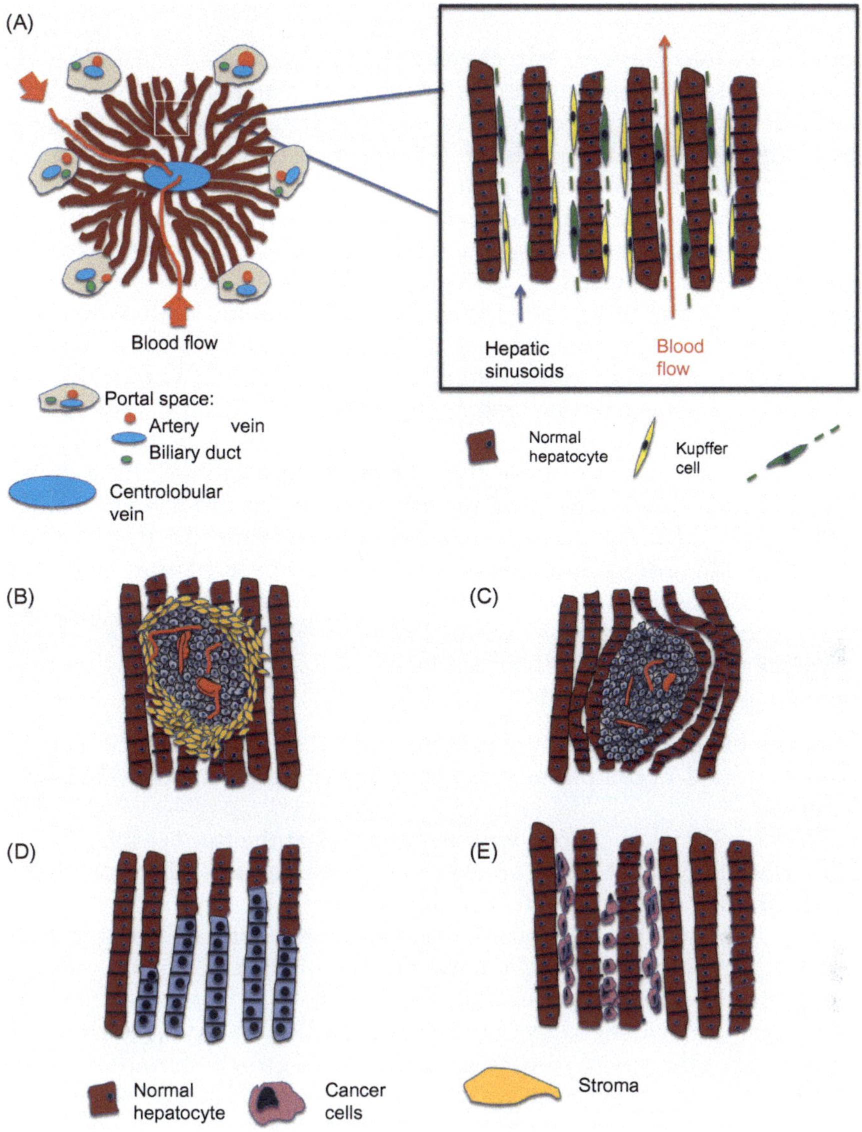

FIGURE 2.2 Vascular patterns in liver tumors. (A) Normal liver: the liver receives both arterial blood (from the hepatic artery) and venous blood (from the portal vein). The smaller branches of both the venous and arterial system are present in the portal spaces. From here the two mix up and flow through the sinusoidal blood vessels, reaching the centrolobular vein. (B) Desmoplastic angiogenic pattern: the neoplastic cells induce angiogenesis and also the formation of a rim of desmoplastic stroma separating the cancer cells from the normal liver. (C) Pushing angiogenic pattern: the neoplastic cells induce angiogenesis; however, no desmoplastic rim is present. The cancer cells "push" away the liver cell plates. (D) Replacement nonangiogenic pattern: observed in both primary and secondary liver tumors, the neoplastic cells "replace" the hepatocytes while the overall organ architecture is preserved. The cancer cell effectively co-opts the sinusoidal blood vessels. (E) Sinusoidal nonangiogenic pattern: this is a rarely seen pattern in which the cells grow inside the lumen of the sinusoids.

patterns is that tumors that express the replacement (nonangiogenic) growth pattern preserve the architecture of the normal liver tissue, while desmoplastic (angiogenic) or pushing (angiogenic) liver metastases do not. The cancer cells of a liver metastasis with replacement HGP infiltrate the liver cell plates, *replace* the hepatocytes, and hijack or co-opt the preexisting sinusoidal blood vessels as a means of vascularization without eliciting angiogenesis. This is in contrast to desmoplastic liver metastases. This type of tumors does not use the surrounding liver tissue but instead elicits a wound healing reaction, creating a rim of fibrosis that separates the cancer cells from the liver parenchyma (Fig. 2.2B). New blood vessels are formed within this *desmoplastic* rim by sprouting angiogenesis. In the pushing growth pattern, which is a rare pattern, the desmoplastic rim is not formed but cancer cells do also not grow into the liver parenchyma. As a consequence, replacement-type liver metastases have an orderly vasculature with minimal distances between blood vessels, while the desmoplastic and the pushing liver metastases have a chaotic vasculature characterized by the so-called vascular hot spots separated by areas with low vessel density.

Vessel co-option in the liver can be recognized by the following morphological characteristics [14]:

- The architecture of the liver is being preserved and the cancer cells are arranged in cell plates in between the co-opted sinusoidal blood vessels.
- Cancer cell makes contact with hepatocytes at the tumor–liver interface and hepatocytes are co-opted by the tumor. Claudins play a role in this heterotypic cell–cell interaction [15].
- The co-opted sinusoidal blood vessels start to express CD34 and lose the expression of LYVE-1, a marker of lymphatic endothelial cells expressed by normal sinusoidal vessels [13].
- Hypoxia (as evidenced by expression of carbonic anhydrase IX) is minimal in the vessel co-opting, replacement growth pattern.
- Deposition of fibrin in the stroma due to vascular leakage is minimal in the vessel co-opting, replacement growth pattern [13].
- There is only minimal inflammation.

The biology of nonangiogenic growth

The biology underlying nonangiogenic tumor growth is still mostly unknown. Furthermore, some aspects of the biology of angiogenic tumors need to be revisited as well.

The aspects investigated so far are mainly related to hypoxia and angiogenic response, inflammation and the immune response, cancer cell motility, cell–cell adhesion, and energy metabolism.

Hypoxia and angiogenesis

Immunohistochemical studies of human non-small-cell lung cancer (NSCLC) have shown that the proteins involved in the VEGF and hypoxia pathways are expressed in angiogenic and nonangiogenic tumors at comparable levels with only a few proteins having a differential expression level (Fig. 2.3) [17,21]. An important exception is thrombospondin which is expressed at significantly higher levels in the stroma of angiogenic cancers [16,21]. The nonangiogenic lung tumors have no desmoplastic stroma and show no stromal thrombospondin expression [17].

Jeong et al. reported thrombospondin expression surrounding vessels in the nonangiogenic lymph node metastases comparably to the normal lymph nodes [20]. In the liver metastases of breast and colorectal cancer, there was clear protein expression of CAIX, as part of the HIF-pathway, in the angiogenic, desmoplastic growth pattern but only minimal expression in the vessel co-opting replacement growth pattern [13]. This coincided with the presence of leaky newly formed blood vessels in the desmoplastic growth pattern, arranged in vascular hot spots due to

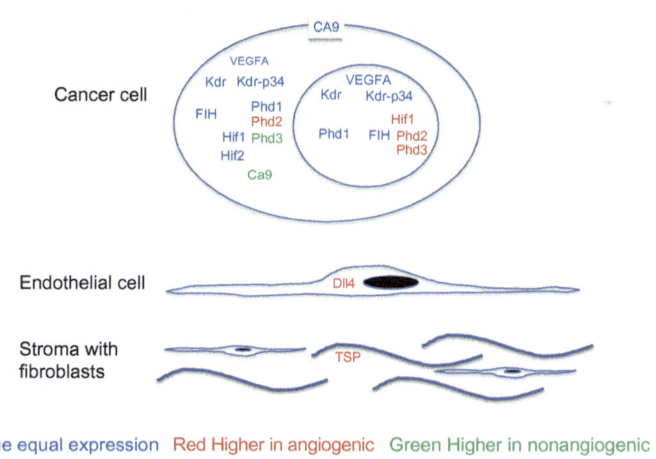

FIGURE 2.3 Angiogenic factors in angiogenic and nonangiogenic tumors. The differences, or lack thereof, in expression and/or transcription of angiogenic factors between angiogenic and nonangiogenic human cancer cells and stroma (endothelial cells and fibroblasts) are shown here. This summary is based on both immunohistochemical and transcriptomics studies of human tumors [16–20].

sprouting angiogenesis, as demonstrated by the presence of fibrin deposits in the stroma. In the replacement growth pattern, the orderly pattern of the co-opted sinusoidal blood vessels of the liver gave rise to much less fibrin depositions.

Two studies have been comparing transcriptomics in angiogenic and nonangiogenic cancer cells. Hu et al. [16] compared nonangiogenic to angiogenic human NSCLC, while Auf et al. [18] compared the U87 glioma cell line WT, which in orthotopic brain implants produces in mice angiogenic tumors, to the U87 dn [with silenced inositol-requiring enzyme (*IRE1*) gene] which instead gave nonangiogenic gliomas. Both studies did not find any differentially expressed hypoxia-related genes and only very few angiogenesis-related genes: *THSB1* in both studies, and *Ang1* and *FGF1* in Auf's work. Auf et al. demonstrated that lower levels of VEGFA transcripts and proteins are present in the IRE1-negative line by qPCR and ELISA [18].

Inflammation and immune response

The angiogenic growth patterns of NSCLC and liver metastases are associated with dense infiltrates of lymphocytes and other immune cells, while the nonangiogenic growth patterns often lack significant inflammation [4,13,17,22]. This suggests that the different histopathological growth patterns have distinct immune phenotypes as defined by Chen and Mellman [23]. The angiogenic growth patterns often display an "excluded" or "inflamed" phenotype, while the nonangiogenic, vessel co-opting growth patterns are mostly "immune deserts."

Both adaptive and innate immune responses can support angiogenesis [24–26] but, concurrently, angiogenesis and the tumor vasculature induce an immunosuppressive phenotype [26,27]. Angiogenic tumors thus seem to co-opt the homeostatic tissue repair program, or wound healing response, combining sprouting angiogenesis and inflammation [27]. In addition, fibrotic factors, such as TGFbeta, and FAP-expressing activated fibroblasts of angiogenic lung and liver metastases may also suppress the antitumor immune response [22]. The inflammatory infiltrate in liver metastases of the desmoplastic type is typically located outside of the metastasis at the interface between the fibrotic capsule and the liver parenchyma [12,13].

In vessel co-opting tumors, other mechanisms that inhibit the immune response may be active. For example, the sinusoidal blood vessels of the liver have highly specialized endothelial cells that scavenge molecules from the blood stream to present these to the hepatocytes. To avoid excessive immune response while performing this action, liver sinusoidal endothelial cells express receptors (such as LSECtin) and immune checkpoints (such as PDL1) that inhibit T-cell activity [28]. So, by co-opting the sinusoidal blood vessels, the replacement-type liver metastases may also acquire an immune suppressive microenvironment.

Motility, invasion, and cell–cell adhesion

Preclinical indications for an increased propensity of vessel co-opting tumors to invade adjacent tissue are corroborated by the clear differences in clinical outcome of patients with this type of tumors when compared with patients with angiogenic tumors. Nonangiogenic NSCLCs have indeed been associated with a higher incidence of metastases during follow-up than when patients had angiogenic lung carcinomas [4,29]. Comparably, in patients operated for colorectal cancer liver metastases, the survival was significantly worse when the liver metastases had a nonangiogenic replacement component, with hazard ratios of 0.4 and 0.5 for overall and relapse-free survival, respectively, in favor of the desmoplastic, angiogenic type of liver metastases [30].

From a histopathological point of view, nonangiogenic lung, liver, and brain tumors have a more irregular interface with the adjacent unaffected tissue than angiogenic tumors [4,12,31]. This suggests that cancer cells, when co-opting the preexisting blood vessels, are more mobile than when a wound healing response with angiogenesis and fibrosis is taking place. In fact, the studies of Barnhill and Lugassy on extravascular migration in melanoma demonstrate that malignant cells, when they respect the microenvironment of the normal tissue, can travel along the blood vessels of the tissue that surrounds a tumor [32], ultimately giving rise to metastases at a distance. Melanoma is, in this respect, a useful model given that neural crest cells, of which melanocytes are an example, undergo the most extensive migration of any embryonic cell type in vertebrate embryos. A characteristic feature of vessel co-opting melanoma liver metastases is the radial extension of individual melanoma cells considerable distances (up to 1 mm) away from the central metastatic focus into the surrounding liver [33].

Several teams have investigated the process of vessel co-option in brain metastases [34–36]. Carbonell et al. have shown that cancer cells of brain metastases co-opt the blood vessels by adhesion to the vascular basement membrane. This "adhesive vessel co-option" relies on beta1-integrins and promotes tumor cell proliferation. Bugyik et al. report comparable results and demonstrate that attachment of cancer cells to the vascular basement membrane promotes survival, tumor growth, and differentiation of cancer cells. The concept of adhesive vessel co-option of brain tumors is corroborated by the studies of Valiente et al. with a role for the axon pathfinding molecule L1CAM as a cell adhesion molecule expressed by cancer cells during vessel co-option. By electron microscopy, Bugyik et al. [36] and Baker et al. [37] have been able to demonstrate the close interaction of the cancer cells and the co-opted blood vessel walls: cancer cells displace the neuropil and detach the astrocytes from the vessels.

Functional evidence of the importance of motility for vessel co-option has been provided in a mouse model of colorectal cancer liver metastasis (HT29) with replacement growth [38]. Arp2/3 mediates the nucleation of actin filaments at the leading edge of cells to drive cell movement. Knockdown of the Arp2/3 subunit ARPC3 by shRNA in the HT29 colorectal cancer cell line suppressed the migration of HT29 cells in vitro, with no potentially confounding effects on cell proliferation, and decreased the replacement HGP in vivo, while significantly increasing the desmoplastic HGP in the HT29 liver metastases [38]. In a mouse model of hepatocellular carcinoma, anti-VEGF treatment by sorafenib led to resistance by co-option of preexisting vessels of the liver and concomitant upregulation of pathways related to cell motility and invasion [39]. Several studies on the inhibition of VEGF yield comparable results: pathways involved in motility and migration of cancer cells are switched on [40]. This effect can be established via-Met-receptor activity [41,42], through epithelial–mesenchymal transition (EMT) [18,41–43], and may involve the WNT, PI3K pathways [44], or axon guidance pathways [19]. A more in-depth discussion on the effect of VEGF, and of VEGF- inhibition, on cancer cell motility can be found in Donnem et al. [45] and Kuczynski et al. [2] (Fig. 2.4).

Energy metabolism

Vessel co-opting tumors may have a reprogrammed energy metabolism. Hu et al. [16] reported higher levels of mRNA transcripts coding for proteins involved in oxidative phosphorylation and mitochondrial biogenesis in nonangiogenic tumors compared with angiogenic tumors. This suggests that a switch from glycolysis to oxidative phosphorylation takes places in vessel co-option. Auf et al. did, however, not observe this change in expression of energy metabolism-related genes in their IRE1-silenced nonangiogenic glioma model [18]. Metabolic reprogramming has been observed when antiangiogenic treatment is applied [46]: hypoxic tumor cells start to produce lactate which is taken up by the nonhypoxic cells to feed the oxidative phosphorylation in a process called metabolic symbiosis.

Acquired resistance to antiangiogenic treatment can often be explained by nonangiogenic tumor progression [38]. Given that this can coincide with metabolic reprogramming, it has been proposed that

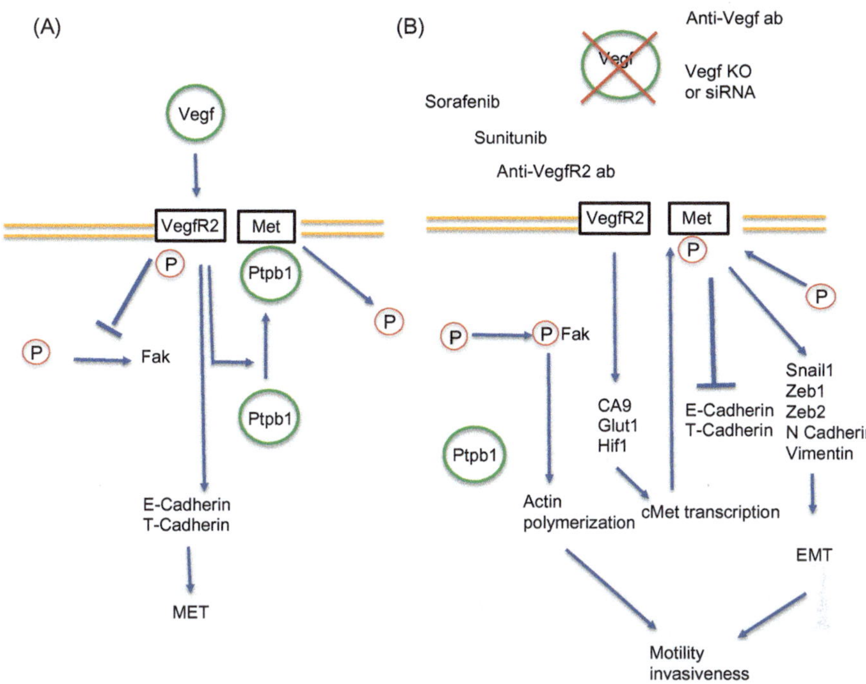

FIGURE 2.4 VEGF inhibition and increased infiltration. One mechanism by which blockage of the VEGF stimulus induces increased motility and invasiveness of tumors has been described by three groups working on glioblastoma [42], pancreatic neuroendocrine tumors [41], and hepatocellular carcinoma [39] models. Their results are consistent and are summarized in this figure. In the presence of VEGF stimulation, there is endothelial proliferation and angiogenesis. At the same time, in the cancer cells, (A) the Ptpb1–protein links to the VegfR2/Met complex inducing dephosphorylation of Met. Phosphorylation of Fak is also inhibited while high levels of E- and T-cadherin are present, leading to MET. When the VEGF-pathway is inhibited, by targeting VEGF or its receptor (B), in addition to a regression of angiogenesis there are also changes in the cancer cells. A group of proteins leading to EMT is upregulated while levels of proteins supporting MET decrease. Hypoxia increases leading to enhanced translation of Met. Fak phosphorylation is also restored leading to increased actine nucleation. As a final result, the cell becomes more mobile and invasive and increased vascular co-option follows (based on Refs. [41,42] and [39]).

targeting the mitochondria of the cancer cells could be a useful strategy in the context of antiangiogenic therapy [47].

What causes tumors to grow in an angiogenic or nonangiogenic, vessel co-opting way?

An important observation related to this question is that the two phenotypes, angiogenesis and vessel co-option, can co-occur within one single tumor, within a patient at a certain time point, and, during the course of the disease [4,6,30,38]. For example, a single liver or lung metastases often has a mixed growth pattern [6,14]. Synchronous colorectal cancer liver metastases can have different growth patterns [48]. Primary nonangiogenic nonsmall cell lung carcinoma can relapse in the brain with angiogenic secondaries [49]. Even clear cell renal cell carcinomas, with angiogenesis induced by a constitutively high HIF and VEGF due to a defective VHL, mainly give rise to vessel co-opting lung metastases [6]. These observations indicate that the growth pattern, and thus the means of vascularization of a tumor, is a highly plastic phenotype. They also suggest that there is an influence of the microenvironment on the growth pattern, although cancer cell-intrinsic factors will be of importance as well in the decision to grow by angiogenesis or vessel co-option. The role of the microenvironment is supported by the following observation. Histological examination revealed that grafting 4T1 (breast cancer) and C26 (colorectal cancer) cell lines in the lungs of mice gave rise to vessel co-opting tumors whereas subcutaneous grafting resulted in angiogenic tumors [6].

Animal model studies have shed some light on the mechanisms involved in this plasticity of vascularization. IRE1, located in the endoplasmic reticulum, is a stress sensor protein that mediates the unfolded protein response. Its role in vascularization has been investigated in a mouse orthotopic brain model and in a chorionic allantoic membrane model of glioblastoma [18]. In both models, the U87 cell line, which expresses the IRE1 protein in its wild-type status, grows as an angiogenic tumor. When, however, the IRE1 gene is silenced, nonangiogenic tumors with a more infiltrative pattern of growth develop. Restoring IL6 levels in IRE1-silenced tumors reinduces angiogenesis while maintaining the aggressive, invasive phenotype. The authors show that IRE1 blocks invasiveness in an IL6-independent way by directly blocking smooth muscle actin (*ACT2*). As far as we know, no studies using human samples have been carried out to investigate the role of IRE1 in determining the means of vascularization.

The former study suggests that the capacity to grow into nearby "normal" tissue of the organ may influence how a tumor obtains its blood vessels. Frentzas et al. have examined the role of actin filament nucleation, one of the important steps to build a functional cytoskeleton. As summarized earlier, knocking down the *ARPC3* gene, coding for one of the components of the Arp2/3 complex, led to decreased motility of the HT29 cell line in vitro and to a switch from nonangiogenic, vessel co-opting liver metastases to angiogenic, desmoplastic liver metastases [38].

Comparable results have been obtained when studying vessel co-option by glioblastoma cells [50]. The Caspani team reports that CDC42, a GTPase regulating the activity of intracellular actin filaments, is involved in the formation of flectopodia which allow for the interaction of the glioblastoma cells with pericytes involving the CD44 adhesion molecule. Blocking of CDC42 prevents vascular co-option by the glioblastoma cells in their model. The authors suggest that this mechanism may play a role in human glioblastoma as they have observed increased CDC42 and CD44 expression in perivascular glioblastoma cells by immunohistochemistry in patient samples.

Carcinoma metastatic to the brain often grows by co-opting the brain capillaries. Valiente and coworkers [35,51] have shown that neuroserpins secreted by the cancer cells may be involved in this process. Plasmin [52] generation in the brain releases soluble Fas-ligand from the membrane of astrcytes, killing the cancer cells by inducing apoptosis. When cancer cells express high levels of neuropserpins, plasminogen is not converted to plasmin, and this apoptotic event does not occur. In addition, L1CAM is not cleaved from the surface of the cancer cells and can be used by the cancer cells to adhere to the blood vessels, which seems to be necessary for vessel co-option.

Other mechanisms of cancer cell–endothelial cell adhesion have been proposed as being necessary for vessel co-option. Connexin 28 and connexin 43 facilitate the formation of functional gap junctions between endothelial cells and metastatic cancer cells in vitro and in several in vivo models of melanoma and breast cancer [53]. Silencing of the respective genes led to the loss of cancer cell–endothelial cell adhesion and the inability to co-opt the preexisting blood vessels. Adhesion between cancer cells and blood vessels seems to be a common theme in several studies on vessel co-option. Carbonell et al. have elegantly demonstrated that malignant cells stick to the basement membrane of blood vessels in the brain using beta-1-integrin. This interaction even leads to redifferentiation of metastatic carcinoma cells [34].

The bradykinin signaling pathway may be involved in "guiding" the cancer cells toward the blood vessels for co-option [54]. For example, glioma cells express bradykinin-2 receptor which is activated by bradykinin produced by endothelial cells. This induces intracellular calcium oscillations which drive the neoplastic cells along the bradykinin gradient toward the blood vessel. Other "angiocrine" (i.e., blood vessel derived) factors may be important for vessel co-option as well and are reminiscent of the interaction between endothelial cells and epithelial progenitor cells during organogenesis in the embryo [55].

Talasila et al. [56] demonstrated that, in murine models, amplified EGFR in human glioblastoma multiforme cells is associated with nonangiogenic behavior, while reversing the amplification is associated with a switch to angiogenesis. Tumor cells were isolated from human gliomas, maintained as spheroids and subsequently implanted in mouse brains. Some spheroids gave rise to angiogenic tumors in the brain, while other spheroids induced vessel co-opting tumors. This coincided with normal copy number of the *EGFR* gene or *EGFR* amplification. About half of the nonangiogenic tumors expressed the EGFRvIII variant. Consequently, treatment of mice with the anti-EGFR antibody cetuximab reduced the size of the nonangiogenic implants only. Lentivirus-induced production of the inactive form of EGFR, downregulating the transcription of the wild-type EGFR, reversed the vessel co-opting tumors to angiogenic ones. The mechanism by which EGFR amplification leads to nonangiogenic growth appears to go by inhibition of the phosphorylation of stat3 and the upregulation of snail and vimentin, suggesting (partial) EMT transition to be involved.

Also, in vessel co-opting hepatocellular carcinomas, gene expression patterns compatible with partial EMT have been observed [39]. In addition, the team of Lugassy and Barnhill have studied vessel co-option (or angiotropism/pericyte mimicry) in melanoma and found gene expression patterns that fit with EMT, melanoma cell motility, and cell–cell adhesion [57].

In conclusion, there is probably not a single factor to be held responsible for the switch from angiogenesis to vessel co-option and vice versa. Common biological themes related to vessel co-option are increased motility and cell plasticity that may be related to partial EMT and the paracrine interaction with the co-opted blood vessels, pericytes, and endothelial cells. This combination of processes is also encountered during the growth of an organ in the embryo with instructive signals from the endothelial cells directing the epithelial progenitor cells to form the organ parenchyma, for example, in the developing liver bud [55].

Blood vessels change when they are co-opted by malignant cells

Blood vessels have been shown to change when co-opted by cancer cells. In nonangiogenic liver metastases, the co-opted sinusoidal vessels lose the expression of LYVE-1, a marker of lymph vessels, and start to express CD34 [13]. This probably coincides with a process called "capillarization" during which co-opted sinusoidal endothelial cells transdifferentiate such that their specialized phenotype, consisting of fenestrations, absent or thin basement membranes, and specialized surface marker expression, is gradually lost.

The morphology of the co-opted blood vessels can also change. In a mouse model of experimental brain metastases, glomeruloid bodies are formed by cancer cells that attach to the basement membrane of the endothelial cells and exert a mechanical pulling force [36,58]. In the glioma mouse model of transient co-option of Holash et al. [59], the first event following co-option of brain capillaries was the upregulation of Angiopoietin 2. In the absence of VEGF, this resulted in vascular regression of the co-opted vessels, hypoxia and subsequent angiogenesis. Küsters et al. show that constitutive overexpression of VEGF-isoforms does not necessarily lead to angiogenic tumors [60]. Instead, the co-opted blood vessels are dilated and become more permeable. This results in even more efficient co-option with increased blood flow and diffusion of nutrients into the tumor.

Conclusion

The study of the biology of nonangiogenic tumors is ongoing and some key characteristics of this type of tumor growth are becoming evident. First, it is now clear that angiogenesis is not a hallmark of cancer. Second, while some tumors are exclusively vascularized by vessel co-option, other tumors have both angiogenic and nonangiogenic areas. Cancer cells can even switch between angiogenic and nonangiogenic means of vascularization. This can depend on the host tissue in which they proliferate or can be in response to treatment. Third, the decision to grow in an angiogenic or nonangiogenic pattern appears to be influenced by the interaction of the cancer cells with the microenvironment. Fourth, nonangiogenic neoplastic cells are associated with increased motility and the ability to infiltrate surrounding tissues. This results in more aggressive tumors with an increased rate of metastatic events. Fifth, nonangiogenic growth is an important mechanism of acquired resistance to antiangiogenic therapy but alternative mechanisms have been described.

However, both the "known unknowns" and, very likely, the "unknown unknowns" still outnumber the above "known knowns" (Donald Rumsfeld, US Department of Defense (DoD), news briefing on February 12, 2002).

For example, we still do not know how tumors "decide" to grow in an angiogenic or nonangiogenic manner. Our understanding of vessel co-option is also limited: as each organ has a unique vascular structure, vessel co-option may rely on organ-specific cancer–endothelial cell interactions. This leads to another unresolved issue: is there a comparable crosstalk between cancer cells and angiogenic blood vessels?

We think that the answers to these questions have the potential to produce novel insights that can result in more efficient treatment of patients with cancer, independent of the means of tumor vascularization.

References

[1] Donnem T, Hu J, Ferguson M, et al. Vessel co-option in primary human tumors and metastases: an obstacle to effective anti-angiogenic treatment? Cancer Med 2013;2:427–36.

[2] Kuczynski EA, Vermeulen PB, Pezzella F, Kerbel RS, Reynolds AR. Vessel co-option in cancer. Nat Rev Clin Oncol 2019; [Epub ahead of print].

[3] Pezzella F, Di Bacco A, Andreola S, Nicholson AG, Pastorino U, Harris AL. Angiogenesis in primary lung cancer and lung secondaries. Eur J Cancer 1996;32A 2494–500.
[4] Pezzella F, Pastorino U, Tagliabue E, et al. Non-small-cell lung carcinoma tumor growth without morphological evidence of neo-angiogenesis. Am J Pathol 1997;151:1417–23.
[5] Szabo V, Bugyik E, Dezso K, et al. Mechanism of tumour vascularization in experimental lung metastases. J Pathol 2015;235:384–96.
[6] Bridgeman VL, Vermeulen PB, Foo S, et al. Vessel co-option is common in human lung metastases and mediates resistance to anti-angiogenic therapy in preclinical lung metastasis models. J Pathol 2017;241:362–74.
[7] Willis RA. The spread of tumours in the human body. London: Churchill; 1934.
[8] Travis WD, Brambilla E, Noguchi M, et al. International association for the study of lung cancer/American thoracic society/European respiratory society international multidisciplinary classification of lung adenocarcinoma. J Thorac Oncol 2011;6 244–85.
[9] Almeida BM, Challacombe SJ, Eveson JW, Smith CG, Leigh IM. A novel lamina lucida component of epithelial and endothelial basement membranes detected by LH39 monoclonal antibody. J Pathol 1992;166:243–53.
[10] Passalidou E, Trivella M, Singh N, et al. Vascular phenotype in angiogenic and non-angiogenic lung non-small cell carcinomas. Br J Cancer 2002;86:244–9.
[11] Sardari Nia P, Colpaert C, Vermeulen P, et al. Different growth patterns of non-small cell lung cancer represent distinct biologic subtypes. Ann Thorac Surg 2008;85 395–405.
[12] Vermeulen PB, Colpaert C, Salgado R, et al. Liver metastases from colorectal adenocarcinomas grow in three patterns with different angiogenesis and desmoplasia. J Pathol 2001;195:336–42.
[13] Stessels F, Van den Eynden G, Van der Auwera I, et al. Breast adenocarcinoma liver metastases, in contrast to colorectal cancer liver metastases, display a non-angiogenic growth pattern that preserves the stroma and lacks hypoxia. Br J Cancer 2004;90:1429–36.
[14] van Dam PJ, van der Stok EP, Teuwen LA, et al. International consensus guidelines for scoring the histopathological growth patterns of liver metastasis. Br J Cancer 2017;117:1427–41.
[15] Tabaries S, McNulty A, Ouellet V, et al. Afadin cooperates with Claudin-2 to promote breast cancer metastasis. Genes Dev 2019;33:180–93.
[16] Hu J, Bianchi F, Ferguson M, et al. Gene expression signature for angiogenic and nonangiogenic non-small-cell lung cancer. Oncogene 2005;24:1212–19.
[17] Ferguson M. Angiogenesis in human lung tumours [DPhil]. Oxford: University of Oxford; 2008.
[18] Auf G, Jabouille A, Guerit S, et al. Inositol-requiring enzyme 1alpha is a key regulator of angiogenesis and invasion in malignant glioma. Proc Natl Acad Sci USA 2010;107:15553–8.
[19] Adighibe O, Leek RD, Fernandez-Mercado M, et al. Why some tumours trigger neovascularisation and others don't: the story thus far. Chin J Cancer 2016;35:18.
[20] Jeong HS, Jones D, Liao S, et al. Investigation of the lack of angiogenesis in the formation of lymph node metastases. J Natl Cancer Inst 2015;107.
[21] Adighibe O. Loss of chaperone protein in human cancer [DPhil]. Oxford: University of Oxford; 2012.
[22] van Dam PJ, Daelemans S, Ross E, et al. Histopathological growth patterns as a candidate biomarker for immunomodulatory therapy. Semin Cancer Biol 2018;52:86–93.

[23] Chen DS, Mellman I. Elements of cancer immunity and the cancer–immune set point. Nature 2017;541:321–30.
[24] de Visser KE, Eichten A, Coussens LM. Paradoxical roles of the immune system during cancer development. Nat Rev Cancer 2006;6:24–37.
[25] Mohr T, Haudek-Prinz V, Slany A, Grillari J, Micksche M, Gerner C. Proteome profiling in IL-1β and VEGFactivated human umbilical vein endothelial cells delineates the interlink between inflammation and angiogenesis. PLoS One 2017;12:e0179065.
[26] Missiaen R, Mazzone M, Bergers G. The reciprocal function and regulation of tumor vessels and immune cells offers new therapeutic opportunities in cancer. Semin Cancer Biol 2018;52(pt 2):107–11.
[27] Motz GT, Coukos G. The parallel lives of angiogenesis and immunosuppression: cancer and other tales. Nat Rev Immunol 2011;11:702–11.
[28] Knolle PA, Gerken G. Local control of the immune response in the liver. Immunol Rev 2000;174:21–34.
[29] Pastorino U, Andreola S, Tagliabue E, et al. Immunocytochemical markers in stage I lung cancer: relevance to prognosis. J Clin Oncol 1997;15:2858–65.
[30] Galjart B, Nierop PMH, van der Stok EP, et al. Angiogenic desmoplastic histopathological growth pattern as a prognostic marker of good outcome in patients with colorectal liver metastases. Angiogenesis 2019;22:355–68.
[31] Berghoff AS, Rajky O, Winkler F, et al. Invasion patterns in brain metastases of solid cancers. Neuro Oncol 2013;15:1664–72.
[32] Lugassy C, Zadran S, Bentolila LA, et al. Angiotropism, pericytic mimicry and extravascular migratory metastasis in melanoma: an alternative to intravascular cancer dissemination. Cancer Microenviron 2014;7:139–52.
[33] Barnhill R, Vermeulen P, Daelemans S, et al. Replacement and desmoplastic histopathological growth patterns: a pilot study of prediction of outcome in patients with uveal melanoma liver metastases. J Pathol Clin Res 2018.
[34] Carbonell WS, Ansorge O, Sibson N, Muschel R. The vascular basement membrane as "soil" in brain metastasis. PLoS One 2009;4:e5857.
[35] Valiente M, Obenauf AC, Jin X, et al. Serpins promote cancer cell survival and vascular co-option in brain metastasis. Cell 2014;156:1002–16.
[36] Bugyik E, Dezso K, Reiniger L, et al. Lack of angiogenesis in experimental brain metastases. J Neuropathol Exp Neurol 2011;70:979–91.
[37] Baker GJ, Yadav VN, Motsch S, et al. Mechanisms of glioma formation: iterative perivascular glioma growth and invasion leads to tumor progression, VEGF-independent vascularization, and resistance to antiangiogenic therapy. Neoplasia 2014;16:543–61.
[38] Frentzas S, Simoneau E, Bridgeman VL, et al. Vessel co-option mediates resistance to anti-angiogenic therapy in liver metastases. Nat Med 2016.
[39] Kuczynski EA, Yin M, Bar-Zion A, et al. Co-option of liver vessels and not sprouting angiogenesis drives acquired sorafenib resistance in hepatocellular carcinoma. J Natl Cancer Inst 2016;108.
[40] Paez-Ribes M, Allen E, Hudock J, et al. Antiangiogenic therapy elicits malignant progression of tumors to increased local invasion and distant metastasis. Cancer Cell 2009;15:220–31.
[41] Sennino B, Ishiguro-Oonuma T, Wei Y, et al. Suppression of tumor invasion and metastasis by concurrent inhibition of c-Met and VEGF signaling in pancreatic neuroendocrine tumors. Cancer Discov 2012;2:270–87.
[42] Lu KV, Chang JP, Parachoniak CA, et al. VEGF inhibits tumor cell invasion and mesenchymal transition through a MET/VEGFR2 complex. Cancer Cell 2012;22:21–35.
[43] Depner C, Zum Buttel H, Böğürcü N, et al. EphrinB2 repression through ZEB2 mediates tumour invasion and anti-angiogenic resistance. Nat Commun 2016;7.

[44] Sakariassen PO, Prestegarden L, Wang J, et al. Angiogenesis-independent tumor growth mediated by stem-like cancer cells. Proc Natl Acad Sci USA 2006;103:16466–71.
[45] Donnem T, Reynolds AR, Kuczynski EA, et al. Non-angiogenic tumours and their influence on cancer biology. Nat Rev Cancer 2018;18:323–36.
[46] Allen E, Missiaen R, Bergeres G. Trimming the vascular tree in tumors: metabolic and immune adaptations. Cold Spring Harb Symp Quant Biol 2016;81:21–9.
[47] Navarro P, Bueno MJ, Zagorac I, et al. Targeting tumor mitochondrial metabolism overcomes resistance to antiangiogenics. Cell Rep 2016;15:2705–18.
[48] Eefsen RL, Van den Eynden GG, Hoyer-Hansen G, et al. Histopathological growth pattern, proteolysis and angiogenesis in chemonaive patients resected for multiple colorectal liver metastases. J Oncol 2012;2012:907971.
[49] Jubb AM, Cesario A, Ferguson M, et al. Vascular phenotypes in primary non-small cell lung carcinomas and matched brain metastases. Br J Cancer 2011;104:1877–81.
[50] Caspani EM, Crossley PH, Redondo-Garcia C, Martinez S. Glioblastoma: a pathogenic crosstalk between tumor cells and pericytes. PLoS One 2014;9:e101402.
[51] Er EE, Valiente M, Ganesh K, et al. Pericyte-like spreading by disseminated cancer cells activates YAP and MRTF for metastatic colonization. Nat Cell Biol 2018;20:966–78.
[52] Weaver A, Weetman AP, Grimm O, et al. Endocrine cancers. In: Kerr DJ, Haller DG, van de Velde CJH, Baumann M, editors. Oxford textbook of oncology. Oxford: Oxford University Press; 2016.
[53] Stoletov K, Strnadel J, Zardouzian E, et al. Role of connexins in metastatic breast cancer and melanoma brain colonization. J Cell Sci 2013;126:904–13.
[54] Montana V, Sontheimer H. Bradykinin promotes the chemotactic invasion of primary brain tumors. J Neurosci 2011;31:4858–67.
[55] Ding BS, Cao Z, Lis R, et al. Divergent angiocrine signals from vascular niche balance liver regeneration and fibrosis. Nature 2014;505:97–102.
[56] Talasila KM, Soentgerath A, Euskirchen P, et al. EGFR wild-type amplification and activation promote invasion and development of glioblastoma independent of angiogenesis. Acta Neuropathol 2013;125:683–98.
[57] Lugassy C, Lazar V, Dessen P, et al. Gene expression profiling of human angiotropic primary melanoma: selection of 15 differentially expressed genes potentially involved in extravascular migratory metastasis. Eur J Cancer 2011;47:1267–75.
[58] Dome B, Timar J, Paku S. A novel concept of glomeruloid body formation in experimental cerebral metastases. J Neuropathol Exp Neurol 2003;62:655–61.
[59] Holash J, Maisonpierre PC, Compton D, et al. Vessel cooption, regression, and growth in tumors mediated by angiopoietins and VEGF. Science 1999;284:1994–8.
[60] Küsters B, Leenders WP, Wesseling P, et al. Vascular endothelial growth factor-A (165) induces progression of melanoma brain metastases without induction of sprouting angiogenesis. Cancer Res 2002;62:341–5.

CHAPTER

3

Vascular co-option

Pedro García-Gómez and Manuel Valiente

Brain Metastasis Group, National Cancer Research Center (CNIO), Madrid, Spain

Vascular co-option in normal and transformed cells

Vascular co-option in cancer was initially described by Holash et al. [1] in gliomas and lung metastasis. The finding reported how tumors remain vascularized without the action of angiogenesis by establishing a physical interaction between cancer cells and pre-existing vessels. This process mimics a similar mechanism developed by a variety of normal cells inhabiting the perivascular niche [2] (Fig. 3.1). A key aspect of vascular co-option is that there is no angiogenesis involved, and thus its molecular regulation follows different mechanisms. Surprisingly, in spite of its potential for treating cancer, specially the disseminated disease, vascular co-option remains poorly understood possibly as the consequence of the intense research dedicated to angiogenesis [3]. However, antiangiogenic treatments have not reached the expectations initially predicted [4–9]. Studies devoted to understand the mechanisms of resistance to antiangiogenic therapy identified the involvement of vascular co-option. This finding confirmed the independence of their molecular regulation, and thus the importance of further characterize the less studied regulatory logic of vascular co-option as an emerging opportunity to impair cancer progression [7–9].

Vascular co-option in non-cancer cells

Noncancer cells interact with the vasculature in multiple diverse scenarios. During the development of nervous system, oligodendrocyte precursor cells (OPCs), which will differentiate into oligodendrocytes to

Tumor Vascularization
DOI: https://doi.org/10.1016/B978-0-12-819494-2.00003-1

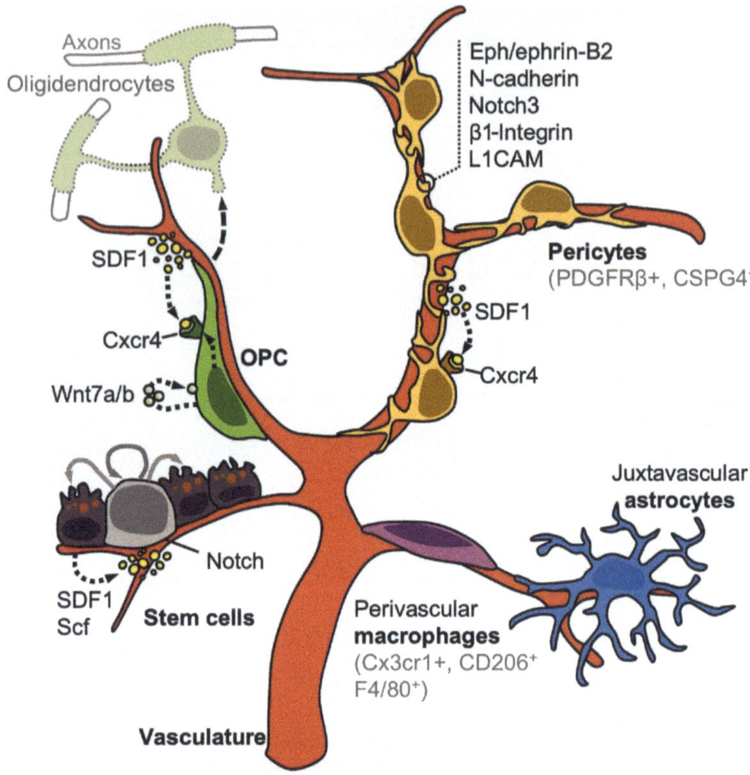

FIGURE 3.1 **Vascular co-option in normal noncancer cells.** Different cell types have been demonstrated to interact with pre-existing vessels. The process of vascular co-option mediates their developmental program, favors their survival, maintains their self-renewal capacity, and could influence the emergence of cell heterogeneity by inducing specific functional properties to co-opting cells. Different regulatory mechanisms in different co-opting cells are indicated. In gray markers used for defining the identity of the specific cell type is shown. *OPC*, oligodendrocyte precursor cell.

produce the myelin covering axons [10], migrate from their origin to their specific target areas using the endothelial surface of neighboring blood vessels [11,12]. This process is regulated by autocrine Wnt signaling. OPCs release Wnt7a and Wnt7b that induce the expression of the chemokine receptor Cxcr4 [12]. Cxcr4 binds to its ligand, the chemokine SDF-1, secreted by co-opted endothelial supporting the migration of OPCs along the vessels [12]. OPCs remain undifferentiated during the migration due to Wnt signaling [12,13] and only mature when they arrive at their final destination, which coincides with the detachment from endothelial cells as they decrease the expression of Wnt pathway and Cxcr4 [12] (Fig. 3.1).

However, the prototypical non-cancer cell performing vascular co-option is the pericyte. Pericytes, the mural cells surrounding endothelial cells of small blood vessels, are required for maintaining vessel stability [14]. In addition, pericytes regulate blood flow due to their contractibility properties and contribute to tissue regeneration given their mesenchymal stem cell properties [14,15]. Pericytes are specially relevant in the brain, where they are key for the homeostasis of the blood–brain barrier (BBB) regulating its maturation during development and then its permeability [14,15]. Not surprisingly, the brain has one of the highest pericyte coverage of the vascular network [14,15]. Pericytes interact with the endothelium using the same mechanism described in OPCs involving the SDF-1/Cxcr4 axis, which is responsible for pericyte recruitment and migration along the vessels [16]. In addition, the interaction between Jagged-1, from endothelial cells, and Notch3, from pericytes, is responsible for vessel stabilization [17]. Eph/ephrin-B2 signaling is also involved in this cell-to-cell interaction [18,19]. Finally, the cell adhesion molecules N-cadherin [20], VCAM1, which binds to endothelial $\alpha 4\beta 1$-integrin activating a downstream signaling influenced by the pericyte proteoglycan NG2 [21,22], and L1CAM, which reinforces the $\beta 1$-integrin/ILK signaling pathway and mediates pericytes spreading and vessel homeostasis [2], are all known mediators of vascular co-option in pericytes (Fig. 3.1).

Other non-cancer cells that are known to reside at the perivascular space are stem cells. For instance, 95% of Hoxb5 + , a recently discovered nuclear marker for hematopoietic stem cells, were detected attached to the endothelial cells in the bone marrow [23]. Similarly, neural stem cells are in direct contact with the special vasculature of the subventricular zone in the brain where the BBB lacks astrocytes endfeets and pericyte coverage [24]. The perivascular location provides stem cells preferential access to molecules from the blood such as nutrients, growth factors, and hormones; as well as angiocrine factors that regulate their differentiation, self-renewal proliferation, and/or migration capacities [24,25]. To mention a few molecular regulators of vascular co-option in stem cells, the SDF-1/Cxcr4 axis is also involved by inducing the expression of $\alpha 6$-integrin to bind to endothelial cell-derived laminin [26,27], and the stem cell ability to self-renew is promoted by Notch signaling [28–30] as well as the production of the Stem Cell Factor (SCF) [31] (Fig. 3.1).

Recent analyses on cell heterogeneity in brain macrophages and astrocytes have described cellular subtypes that have been mainly defined by their perivascular location [32,33]. Although the molecular nature of their interaction with the vessels is unknown, existing evidence suggests that this specific location could be a major contributor to their functional specialization, thus participating in the emergence of cell heterogeneity [33] (Fig. 3.1).

Vascular co-option in cancer cells

Vascular co-option has been described not only in experimental models but also in patients affected by glioma [1,34–36], melanoma [37–39], lung cancer [4,9,40–42], breast cancer [43–46], renal cancer [47], and liver cancer [7,8] both in the primary tumor [1,8,34–37,40] and in distant organs such as brain [2,38,40,43,44,46,48–54], lungs [1,2,9,37,42,44], bone [2,44,55], or liver [2,7] where metastatic cells disseminate (Fig. 3.2A).

In primary tumors, cancer cells in close contact with vessels receive and have access to oxygen, nutrients, and other factors produced by endothelial cells (angiocrine factors) that support their viability and growth [56,57]. This is particularly important to cancer stem cells (CSCs), which can maintain their properties by getting access to secreted angiocrine factors such as vascular endothelial growth factor (VEGF), which promotes their survival [58,59], Notch ligands like Jagged-1, which activates CSCs nurturing their self-renewal capacity, chemoresistance, and tumorigenicity [60,61]. As the primary tumor develops, the ability of cancer cells to co-opt the vessels initiates the process by which few of them will get access to systemic circulation upon intravasation, thus starting the metastatic cascade [37].

The interaction between cancer cells and vessels is resumed once metastatic cells reach the capillary network of distal organs after they disseminate from the primary tumor through systemic circulation. Interestingly, those tumor cells that extravasate do not abandon the perivascular niche but remain closely attached to the abluminal side of the vessels performing vascular co-option [50,53]. The initial observation of this process in lung metastasis [1] has been extended to multiorgan metastases, establishing vascular co-option as a hallmark of metastasis-initiating cells [2,38,41,46,50,52,53]. Besides the benefits described in the primary tumor [56,57], there are additional implications of vascular co-option involving several cancer hallmarks [62] such as aggressive growth, immune evasion, latency, and resistance to therapy [8,9,35,37,44,49,55,63,64] that are especially relevant in metastasis.

Molecular regulation of vascular co-option

Given that vascular co-option has been linked to several cancer hallmarks, its molecular characterization is key to develop novel anticancer strategies, which might have relevant implications to prevent the development of metastasis.

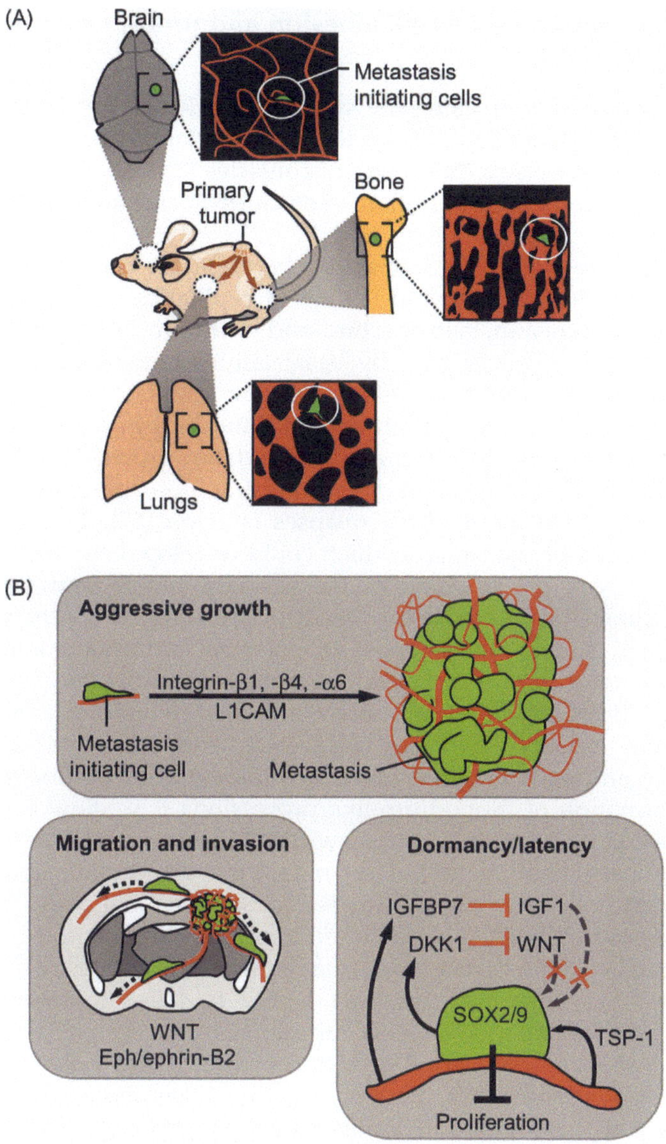

FIGURE 3.2 **Implications of vascular co-option in cancer.** (A) Once they arrive to a secondary organ, disseminated metastasis-initiating cells remain attached to the abluminal side of vessels after extravasation performing vascular co-option. This process has been described in multiple organs targeted by metastasis and shown to be independent of the primary tumor type. (B) Vascular co-option favors many aspects that are highly relevant for metastasis such as the development of aggressive growth, the ability to invade locally, and the ability to remain alive but dormant over long periods of time.

Molecular regulation of cell adhesion and proliferation in vascular co-option

Immediately after crossing the vascular barrier of any given secondary organ, metastasis-initiating cells remain attached to the abluminal side of the vessels using preexisting capillaries [2,46,50,53]. This interaction is dependent on adhesion molecules including integrins and L1CAM [50,52].

β1-Integrin mediates the interaction between the cancer cell and different components of the basal lamina of capillaries including collagen I and IV, fibronectin, vitronectin, and laminin [52]. Activation of β1-integrin in the cancer cells by basal lamina components initiates a signaling cascade involving FAK-dependent ERK1/2 phosphorylation, which supports the proliferation of metastatic cells in secondary organs upon arrival [52] (Fig. 3.2B). In addition to favor growth, this molecular mechanism is also important for protecting not aggressive dormant cancer cells, which might drive relapses later on [55]. This scenario is very relevant in breast cancer, which could develop a significant period of time between the removal of the primary tumor until metastases manifest clinically. This clinical scenario is explained by the ability of disseminated tumor cells (DTCs) to enter in a dormant state in the secondary organ until they found favorable conditions to grow [65,66]. DTCs in a dormant state have been shown to remain attached to the perivascular niche [44] being protected from chemotherapy in a β1-integrin and $\alpha_V\beta_3$ integrin-dependent manner [55]. β1-Integrin is key for the formation of filopodium-like protrusions that allow cancer cells to established cell–cell interactions with components of the extracellular matrix of basal lamina [67]. The underlying molecular regulation of these protrusions involves Rif/mDia2 actin machinery and the ILK/β-parvin/cofilin signaling pathway that signals through the FAK–ERK pathway [67]. Thus vascular co-option benefit cancer cells in a β1-integrin-dependent manner by granting access to the components of the basal lamina of co-opted vessels. In addition, cancer cells might also benefit by the influence of β1-integrin on the production of angiocrine factors in endothelial cells. This process has been shown to be dependent on β1-integrin, given its ability to couple growth with blood flow leading to the activation of mechanotransduction processes that induces the secretion of these factors [56].

Additional integrins have been described to participate in vascular co-option, which suggests certain plasticity within the communication between cancer cells and the endothelium. For instance, acute lymphoblastic leukemia cancer cells metastasize to the central nervous system without the need to cross the BBB, given their ability to exploit arachnoid veins by α6-integrin, which allows them to interact with laminin

[68]. Binding of β4-integrin from cancer cells to basal lamina components has been shown to activate ErbB2 by the ability of the integrin cytoplasmic domain to activate receptor tyrosine kinases (RTKs). This signaling pathway that links vascular co-option to the activation of RTKs stimulates the production of secreted molecules, such as VEGF, that will modify endothelial cells to favor tumor growth [43].

Besides integrins, additional cell adhesion molecules are involved in vascular co-option. L1CAM, which is expressed during the development of the nervous system [69], becomes deregulated in cancer and correlates with poor prognosis in many tumor types [47,70]. This cell adhesion molecule was initially found to be enriched at the interface between tumor and endothelial cell in brain metastasis [50] and later involved multiorgan metastases from different primary tumors [2]. Loss of function of L1CAM in cancer cells mimics the phenotype of targeting β1-integrin, which involves the inability of cancer cells to spread on co-opted vessels thus blocking their proliferation and consequently preventing the development of macrometastasis [50]. L1CAM promotes β1-integrin signaling by increasing PAK1/2 phosphorylation in an ILK-dependent manner. PAK1/2 phosphorylation promotes the formation of actin filaments in an ankirin2-dependent manner, which allows the formation of cell protrusions required for cell spreading on vessels [2], which might alternatively involve Arp2/3 [2]. The induction of cancer cell spreading along the preexisting vessels activates YAP pathway, which induces a gene expression signature responsible for reactivating the proliferative program when metastatic cells initiate the colonization of a new organ [2]. Thus, targeting the integrin–L1CAM signaling pathway in cancer cells impairs the initial stages of metastasis colonization and prevents the development of macrometastases as has been shown in different experimental cancer models including lung, breast, skin, colorectal, and renal cancer as well as lymphoma [2,50,52,55].

Although the process of vascular co-option involves the cancer cell and the endothelial cell, it could be influenced by additional cells of the microenvironment induced by inflammation. This has been demonstrated in melanoma, a type of skin cancer highly linked to UV radiation, where vascular co-option has been shown to be dependent on HMGB1–TLR4–dependent inflammation [37]. Excessive UV light induces the secretion of the danger-associated molecular pattern HMGB1 from damaged melanocytes. HMGB1 is detected by neutrophils through the TLR4 receptor producing an inflammatory response that favors the ability of melanoma cells to co-opt the vessels at the primary tumor and then disseminate systemically [37]. The enhanced co-option of cancer cells seems to be dependent on the TNF released by the neutrophils during the inflammatory response [37].

Molecular regulation of migration and invasion linked to vascular co-option

The process of migration is highly relevant in cancer. Dissemination of cancer cells out of the primary tumor is a necessary step for the development of systemic cancer but also local dissemination can have important clinical implications. For instance, intracerebral invasion of glioma cells is a marker of poor prognosis [71]. Vascular co-option is highly relevant in the ability of cancer cells to migrate in the brain. Glioma cells exploit ephrin signaling to move along pre-existing capillaries [35]. Normally, Eph/ephrin signaling constrains the migration of incipient premalignant lesions [72]; however, as the tumor evolves and increases its aggressiveness this initial antitumor mechanism gets rewired. Invasive cancer cells upregulate ephrin-B2 levels to activate Eph forward signaling through homotypic cell–cell interactions that support migration [35] (Fig. 3.2B).

The acquisition of migratory properties in glioma cells could also be linked to the cell of origin [36]. Glioma cells with an Olig2 + oligodendrocyte precursor origin, which are cells performing vascular co-option during their normal development [12], are more prone to invade using the preexisting vessels [36]. Glioma cells hijack the developmental pathway used by OPC to migrate, which involves the secretion of Wnt7b (Fig. 3.1). Besides the poorer prognosis of infiltrative gliomas, the fact that vascular co-option maintains the structure of the BBB better than the disruption induced in the process of angiogenesis could have important implications for treatment [73]. In spite of being highly sensitive to Wnt inhibitors such as porcupine, Olig2 + glioma cells are more protected from chemotherapies that do not cross the BBB than Olig2 − cells. In contrast, the infiltration of immune cells will be higher in angiogenic Olig2 − tumors than Olig2 + gliomas. This will define the type of immunophenotype that these tumor types will develop [36], which might have important implications in emerging strategies using the immune system to target cancer cells.

Molecular regulation of dormancy, latency, and awakening linked to vascular co-option

As slow-proliferative normal stem cells located at the perivascular niche do, cancer cells could develop a state of quiescence associated with vessels. This state could be induced by molecules produced by endothelial cells or derived from the intrinsic properties of cancer cells. Although several programs have been described, it is not known whether they are interconnected.

Thrombospondin-1 (TSP-1) was initially described as a molecule secreted by endothelial cells capable of preventing tumor growth due to the inhibition of angiogenesis [74,75]. Additionally, in brain, bone, and lung metastasis, co-opted endothelial cells producing TSP-1 induce dormancy on co-opting cancer cells through unknown mechanism (Fig. 3.2). Complementarily, endothelial cells undergoing sprouting angiogenesis downregulate TSP-1 and increase TGFβ1 and periostin that promotes tumor cell growth [44,76].

Additional programs of latency described in metastatic cells at the perivascular space depend on the presence of the transcription factors SOX2 and SOX9, which induce the expression of the Wnt inhibitor DKK1. Cancer cell–secreted DKK1 prevents Wnt-dependent influence on B-catenin, thus blocking its potential positive influence on proliferation [49]. Thus when DKK1 is targeted in cancer cells, metastatic cells transit out of dormancy and resume proliferation. However, cancer cells forced to leave the latent state are rapidly eliminated by host innate defenses due to the induction of NK-cells ligands expression derived from loss of SOX expression and gain of Wnt signaling [49] (Fig. 3.2). Consequently, the switch from dormant/latent state to a productive proliferative state requires additional regulatory mechanisms. Metastatic cells from breast cancer in the lung, bone, and brain use TM4SF1 to switch from dormancy to aggressive growth [63]. This phenotypic switch in vascular co-opting cells is initiated by the upregulation of TM4SF1, which couples the collagen I receptor DDR1 to PKCα that activates the JAK–STAT pathway, which effectively drives proliferation of metastatic cells [63]. Interestingly, this non-canonical TM4SF1 signaling pathway also enhances CSC properties inducing SOX2 and NANOG, which reinforces the link of CSC traits and metastatic reactivation [63]. Additional mechanisms developed by cancer cells to escape from the proliferative-checkpoint imposed at the perivascular niche have been described. By inducing the secretion of FGF4, indolent tumor cells co-opting blood vessels are able to activate the FGFR1 in endothelial cells, which drives ETS2-dependent downregulation of IGFBP7 expression [64]. Consequently, co-opting cancer cells start to receive increased amounts of IGF1 due to the reduction of extracellular IGFBP7, which was previously sequestering the growth factor and preventing its strong effect on proliferation [64] (Fig. 3.2). IGF1 not only promotes tumor aggressiveness but also makes cancer cells chemoresistant [64].

Preclinical applications to target vascular co-option: prevention of metastasis

Genetic targeting of cancer cell adhesion molecules L1CAM and β1-integrin has probed the potential of developing anti-metastatic pharmacological strategies on the basis of vascular co-option. Both spontaneous and induced metastases from lung cancer, breast cancer, renal cancer and melanoma potentially affecting the bone, lungs, brain, and liver were dramatically reduced after targeting any of these cell adhesion molecules [2,39,50,52]. Not only metastases but also difficult-to-treat and life-threatening gliomas were highly sensitive to a combination therapy based on an antiangiogenic drug (bevacizumab) and a β1-integrin antagonist (OS2966) when applied to xenografts [77]. A similar combination therapy based on capecitabine, an antiangiogenic drug (B20-4.1.1), and the genetic inhibition of the Arp2/3 complex, required for vascular co-option, was superior to anti-angiogenesis monotherapy when applied to experimental breast cancer models that develop spontaneous liver metastasis [7]. Additional successful therapeutic strategies targeted Ephrin-B2 using blocking antibodies in experimental glioblastoma models [35].

Besides blocking vascular co-option to target metastasis, impairing the support provided by the perivascular niche has been shown to reduce the increased resistance to chemotherapies. For instance, bone metastasis were sensitized to doxorubicin and taxol by using blocking antibodies against β1 and $\alpha_V\beta_3$ [55]. Consequently, using therapeutic strategies targeting vascular co-option might be useful to reduce resistant cells at the perivascular niche that will drive relapse later on. Similarly, cells that might escape surgery because they have migrated out of the tumor core is a frequent cause of relapse in gliomas. Targeting cancer cells co-opting vessels by combining Wnt inhibitors LGK974 or XAV939 with the standard of care temozolomide increased the control of Olig2+ gliomas demonstrating the value of this adjuvant therapy [36].

One important caveat of targeting vascular co-option as an anticancer strategy is the potential toxic effects of targeting normal cells that require a perivascular location (Fig. 3.1). Although several examples of normal cells developing vascular co-option are specific to development (i.e., L1CAM for axonal growth), others must be considered (i.e., pericytes or adult stem cells) since damaging them will have important implications. Consequently, besides genetic strategies that have been a valuable proof-of-concept to demonstrate the enormous potential of targeting vascular co-option as a novel anti-metastasis strategy, more pharmacological strategies are needed specially directed against aspects of vascular co-option specific for cancer cells.

Acknowledgments

Research in the Brain Metastasis Group is supported by MINECO-Retos SAF2017-89643-R (M.V.), Bristol-Myers Squibb-Melanoma Research Alliance Young Investigator Award (498103) (M.V.), Beug Foundation's Prize for Metastasis Research (M.V.), Fundación Ramón Areces (CIVP19S8163) (M.V.), Worldwide Cancer Research (19-0177) (M.V.), H2020-FETOPEN (828972) (M.V.), CLIP Award Cancer Research Institute (54545) (M.V.), AECC Coordinated Translational Groups 2017 (GCTRA16015SEOA) (M.V.). P.G-G. is the recipient of La Caixa Foundation (ID100010434) and European Union's Horizon 2020 research and innovation programme under the Marie Sklodowska-Curie grant agreement (No. 713673) PhD Program Fellowship (LCF/BQ/IN17/11620028). M.V. is a Ramón y Cajal Investigator (RYC-2013-13365) and EMBO YIP (4053).

References

[1] Holash J, Maisonpierre PC, Compton D, Boland P, Alexander CR, Zagzag D, et al. Vessel cooption, regression, and growth in tumors mediated by angiopoietins and VEGF. Science 1999;284(5422):1994–8.

[2] Er EE, Valiente M, Ganesh K, Zou Y, Agrawal S, Hu J, et al. Pericyte-like spreading by disseminated cancer cells activates YAP and MRTF for metastatic colonization. Nat Cell Biol 2018;20:966–78.

[3] Jain RK. Antiangiogenesis strategies revisited: from starving tumors to alleviating hypoxia. Cancer Cell 2014;26(5):605–22.

[4] Ilhan-Mutlu A, Osswald M, Liao Y, Gömmel M, Reck M, Miles D, et al. Bevacizumab prevents brain metastases formation in lung adenocarcinoma. Mol Cancer Ther 2016;15(4):702–10.

[5] Ilhan-Mutlu A, Siehs C, Berghoff AS, Ricken G, Widhalm G, Wagner L, et al. Expression profiling of angiogenesis-related genes in brain metastases of lung cancer and melanoma. Tumor Biol 2016;37(1):1173–82.

[6] Yano S, Shinohara H, Herbst RS, Kuniyasu H, Bucana CD, Ellis LM, et al. Expression of vascular endothelial growth factor is necessary but not sufficient for production and growth of brain metastasis. Cancer Res [Internet] 2000;60(17):4959–67. Available from: <http://cancerres.aacrjournals.org/content/canres/60/17/4959.full.pdf>.

[7] Frentzas S, Simoneau E, Bridgeman VL, Vermeulen PB, Foo S, Kostaras E, et al. Vessel co-option mediates resistance to anti-angiogenic therapy in liver metastases. Nat Med [Internet] 2016;22:1294. Available from: https://doi.org/10.1038/nm.4197.

[8] Reynolds AR, Bar-Zion A, Lee CR, Kuczynski EA, Foster FS, Daley F, et al. Co-option of liver vessels and not sprouting angiogenesis drives acquired sorafenib resistance in hepatocellular carcinoma. JNCI J Natl Cancer Inst [Internet] 2016;108(8). Available from: https://dx.doi.org/10.1093/jnci/djw030.

[9] Bridgeman VL, Vermeulen PB, Foo S, Bilecz A, Daley F, Kostaras E, et al. Vessel co-option is common in human lung metastases and mediates resistance to anti-angiogenic therapy in preclinical lung metastasis models. J Pathol [Internet] 2017;241 (3):362–74. Available from: https://doi.org/10.1002/path.4845.

[10] Fünfschilling U, Supplie LM, Mahad D, Boretius S, Saab AS, Edgar J, et al. Glycolytic oligodendrocytes maintain myelin and long-term axonal integrity. Nature 2012;485 (7399):517.

[11] Kessaris N, Fogarty M, Iannarelli P, Grist M, Wegner M, Richardson WD. Competing waves of oligodendrocytes in the forebrain and postnatal elimination of an embryonic lineage. Nat Neurosci 2006;9(2):173.

[12] Tsai H-H, Niu J, Munji R, Davalos D, Chang J, Zhang H, et al. Oligodendrocyte precursors migrate along vasculature in the developing nervous system. Science [Internet] 2016;351(6271):379–84. Available from: <http://science.sciencemag.org/content/351/6271/379.abstract>.

[13] Fancy SPJ, Baranzini SE, Zhao C, Yuk D-I, Irvine K-A, Kaing S, et al. Dysregulation of the Wnt pathway inhibits timely myelination and remyelination in the mammalian CNS. Genes Dev 2009;23(13):1571–85.

[14] Armulik A, Genové G, Betsholtz C. Pericytes: developmental, physiological, and pathological perspectives, problems, and promises. Dev Cell [Internet] 2011;21 (2):193–215. Available from: <http://www.sciencedirect.com/science/article/pii/S1534580711002693>.

[15] Obermeier B, Daneman R, Ransohoff RM. Development, maintenance and disruption of the blood-brain barrier. Nat Med 2013;19(12):1584.

[16] Song N, Huang Y, Shi H, Yuan S, Ding Y, Song X, et al. Overexpression of platelet-derived growth factor-BB increases tumor pericyte content via stromal-derived factor-1α/CXCR4 axis. Cancer Res [Internet] 2009;69(15):6057–64. Available from: <http://cancerres.aacrjournals.org/content/69/15/6057>.

[17] Hua L, Simone K, Brenda L. NOTCH3 expression is induced in mural cells through an autoregulatory loop that requires endothelial-expressed JAGGED1. Circ Res [Internet] 2009;104(4):466–75. Available from: https://doi.org/10.1161/CIRCRESAHA.108.184846.

[18] Foo SS, Turner CJ, Adams S, Compagni A, Aubyn D, Kogata N, et al. Ephrin-B2 controls cell motility and adhesion during blood-vessel-wall assembly. Cell [Internet] 2006;124(1):161–73. Available from: <http://www.sciencedirect.com/science/article/pii/S0092867405012341>.

[19] Salvucci O, Maric D, Economopoulou M, Sakakibara S, Merlin S, Follenzi A, et al. EphrinB reverse signaling contributes to endothelial and mural cell assembly into vascular structures. Blood [Internet] 2009;114(8):1707–16. Available from: <http://www.bloodjournal.org/content/114/8/1707>.

[20] Gerhardt H, Wolburg H, Redies C. N-cadherin mediates pericytic-endothelial interaction during brain angiogenesis in the chicken. Dev Dyn [Internet] 2000;218(3):472–9. Available from: https://doi.org/10.1002/1097-0177(200007)218:3%3C472::AID-DVDY1008%3E3.0.CO.

[21] You W-K, Yotsumoto F, Sakimura K, Adams RH, Stallcup WB. NG2 proteoglycan promotes tumor vascularization via integrin-dependent effects on pericyte function. Angiogenesis [Internet] 2014;17(1):61–76. Available from: https://doi.org/10.1007/s10456-013-9378-1.

[22] Garmy-Susini B, Jin H, Zhu Y, Sung R-J, Hwang R, Varner J. Integrin α4β1–VCAM-1–mediated adhesion between endothelial and mural cells is required for blood vessel maturation. J Clin Invest [Internet] 2005;115(6):1542–51. Available from: https://doi.org/10.1172/JCI23445.

[23] Chen JY, Miyanishi M, Wang SK, Yamazaki S, Sinha R, Kao KS, et al. Hoxb5 marks long-term haematopoietic stem cells and reveals a homogenous perivascular niche. Nature [Internet] 2016;530:223. Available from: https://doi.org/10.1038/nature16943.

[24] Tavazoie M, Van der Veken L, Silva-Vargas V, Louissaint M, Colonna L, Zaidi B, et al. A specialized vascular niche for adult neural stem cells. Cell Stem Cell [Internet] 2008;3(3):279–88. Available from: <http://www.sciencedirect.com/science/article/pii/S1934590908003962>.

[25] Goldberg JS, Hirschi KK. Diverse roles of the vasculature within the neural stem cell niche. Regen Med [Internet] 2009;4(6):879–97. Available from: https://doi.org/10.2217/rme.09.61.

[26] Shen Q, Wang Y, Kokovay E, Lin G, Chuang S-M, Goderie SK, et al. Adult SVZ stem cells lie in a vascular niche: a quantitative analysis of niche cell-cell interactions. Cell Stem Cell [Internet] 2008;3(3):289–300. Available from: <http://www.sciencedirect.com/science/article/pii/S1934590908003974>.

[27] Kokovay E, Goderie S, Wang Y, Lotz S, Lin G, Sun Y, et al. Adult SVZ lineage cells home to and leave the vascular niche via differential responses to SDF1/CXCR4 signaling. Cell Stem Cell [Internet] 2010;7(2):163–73. Available from: <http://www.sciencedirect.com/science/article/pii/S1934590910002791>.

[28] Hitoshi S, Alexson T, Tropepe V, Donoviel D, Elia AJ, Nye JS, et al. Notch pathway molecules are essential for the maintenance, but not the generation, of mammalian neural stem cells. Genes Dev 2002;16(7):846–58.

[29] Nakamura Y, Sakakibara S, Miyata T, Ogawa M, Shimazaki T, Weiss S, et al. The bHLH gene Hes1 as a repressor of the neuronal commitment of CNS stem cells.

J Neurosci [Internet] 2000;20(1):283–93. Available from: <http://www.jneurosci.org/content/20/1/283>.

[30] Shen Q, Goderie SK, Jin L, Karanth N, Sun Y, Abramova N, et al. Endothelial cells stimulate self-renewal and expand neurogenesis of neural stem cells. Science [Internet] 2004;304(5675):1338–40. Available from: <https://science.sciencemag.org/content/304/5675/1338>.

[31] Ding L, Saunders TL, Enikolopov G, Morrison SJ. Endothelial and perivascular cells maintain haematopoietic stem cells. Nature [Internet] 2012;481:457. Available from: https://doi.org/10.1038/nature10783.

[32] Goldmann T, Wieghofer P, Jordão MJC, Prutek F, Hagemeyer N, Frenzel K, et al. Origin, fate and dynamics of macrophages at central nervous system interfaces. Nat Immunol [Internet] 2016;17:797. Available from: https://doi.org/10.1038/ni.3423.

[33] Bardehle S, Krüger M, Buggenthin F, Schwausch J, Ninkovic J, Clevers H, et al. Live imaging of astrocyte responses to acute injury reveals selective juxtavascular proliferation. Nat Neurosci [Internet] 2013;16:580. Available from: https://doi.org/10.1038/nn.3371.

[34] Cai Y, Wu J, Li Z, Long Q. Mathematical modelling of a brain tumour initiation and early development: a coupled model of glioblastoma growth, pre-existing vessel co-option, angiogenesis and blood perfusion. PLoS One [Internet] 2016;11(3):e0150296. Available from: https://doi.org/10.1371/journal.pone.0150296.

[35] Krusche B, Ottone C, Clements MP, Johnstone ER, Goetsch K, Lieven H, et al. EphrinB2 drives perivascular invasion and proliferation of glioblastoma stem-like cells. Elife [Internet], 5. 2016. p. e14845. Gilbertson RJ, editor. https://doi.org/10.7554/eLife.14845.

[36] Griveau A, Seano G, Shelton SJ, Kupp R, Jahangiri A, Obernier K, et al. A glial signature and Wnt7 signaling regulate glioma-vascular interactions and tumor microenvironment. Cancer Cell [Internet] 2018;33(5) 874–89.e7. Available from: <http://www.sciencedirect.com/science/article/pii/S1535610818301259>.

[37] Bald T, Quast T, Landsberg J, Rogava M, Glodde N, Lopez-Ramos D, et al. Ultraviolet-radiation-induced inflammation promotes angiotropism and metastasis in melanoma. Nature [Internet] 2014;507:109. Available from: https://doi.org/10.1038/nature13111.

[38] Bentolila LA, Prakash R, Mihic-Probst D, Wadehra M, Kleinman HK, Carmichael TS, et al. Imaging of angiotropism/vascular co-option in a murine model of brain melanoma: implications for melanoma progression along extravascular pathways. Sci Rep 2016;6:23834.

[39] Ernst A-K, Putscher A, Samatov TR, Suling A, Galatenko VV, Shkurnikov MY, et al. Knockdown of L1CAM significantly reduces metastasis in a xenograft model of human melanoma: L1CAM is a potential target for anti-melanoma therapy. PLoS One [Internet] 2018;13(2):e0192525. Available from: https://doi.org/10.1371/journal.pone.0192525.

[40] Jubb AM, Cesario A, Ferguson M, Congedo MT, Gatter KC, Lococo F, et al. Vascular phenotypes in primary non-small cell lung carcinomas and matched brain metastases. Br J Cancer [Internet] 2011;104:1877. Available from: http://dx.doi.org/10.1038/bjc.2011.147.

[41] Nguyen DX, Chiang AC, Zhang XH-F, Kim JY, Kris MG, Ladanyi M, et al. WNT/TCF signaling through LEF1 and HOXB9 mediates lung adenocarcinoma metastasis. Cell 2009;138(1):51–62.

[42] Shibue T, Weinberg RA. Integrin β1-focal adhesion kinase signaling directs the proliferation of metastatic cancer cells disseminated in the lungs. Proc Natl Acad Sci 2009; pnas. 0904227106.

[43] Fan J, Cai B, Zeng M, Hao Y, Giancotti FG, Fu BM. Integrin β4 signaling promotes mammary tumor cell adhesion to brain microvascular endothelium by inducing ErbB2-mediated secretion of VEGF. Ann Biomed Eng 2011;39(8):2223–41.
[44] Ghajar CM, Peinado H, Mori H, Matei IR, Evason KJ, Brazier H, et al. The perivascular niche regulates breast tumour dormancy. Nat Cell Biol 2013;15(7):807.
[45] Lockman PR, Mittapalli RK, Taskar KS, Rudraraju V, Gril B, Bohn KA, et al. Heterogeneous blood-tumor barrier permeability determines drug efficacy in mouse brain metastases of breast cancer. Clin Cancer Res 2010; clincanres. 1564.2010.
[46] Lorger M, Felding-Habermann B. Capturing changes in the brain microenvironment during initial steps of breast cancer brain metastasis. Am J Pathol 2010;176 (6):2958–71.
[47] Doberstein K, Wieland A, Lee SBB, Blaheta RAA, Wedel S, Moch H, et al. L1-CAM expression in ccRCC correlates with shorter patients survival times and confers chemoresistance in renal cell carcinoma cells. Carcinogenesis 2010;32(3):262–70.
[48] Lorger M, Krueger JS, O'Neal M, Staflin K, Felding-Habermann B. Activation of tumor cell integrin αvβ3 controls angiogenesis and metastatic growth in the brain. Proc Natl Acad Sci 2009;106(26):10666–71.
[49] Malladi S, Macalinao DG, Jin X, He L, Basnet H, Zou Y, et al. Metastatic latency and immune evasion through autocrine inhibition of WNT. Cell 2016;165(1):45–60.
[50] Valiente M, Obenauf AC, Jin X, Chen Q, Zhang XH-F, Lee DJ, et al. Serpins promote cancer cell survival and vascular co-option in brain metastasis. Cell 2014;156 (5):1002–16.
[51] Berghoff AS, Rajky O, Winkler F, Bartsch R, Furtner J, Hainfellner JA, et al. Invasion patterns in brain metastases of solid cancers. Neuro Oncol 2013;15(12):1664–72.
[52] Carbonell WS, Ansorge O, Sibson N, Muschel R. The vascular basement membrane as "soil" in brain metastasis. PLoS One 2009;4(6):e5857.
[53] Kienast Y, Von Baumgarten L, Fuhrmann M, Klinkert WEF, Goldbrunner R, Herms J, et al. Real-time imaging reveals the single steps of brain metastasis formation. Nat Med 2010;16(1):116.
[54] Kim LS, Huang S, Lu W, Lev DC, Price JE. Vascular endothelial growth factor expression promotes the growth of breast cancer brain metastases in nude mice. Clin Exp Metastasis 2004;21(2):107–18.
[55] Carlson P, Dasgupta A, Grzelak CA, Kim J, Barrett A, Coleman IM, et al. Targeting the perivascular niche sensitizes disseminated tumour cells to chemotherapy. Nat Cell Biol [Internet] 2019;21(2):238–50. Available from: https://doi.org/10.1038/s41556-018-0267-0.
[56] Lorenz L, Axnick J, Buschmann T, Henning C, Urner S, Fang S, et al. Mechanosensing by β1 integrin induces angiocrine signals for liver growth and survival. Nature [Internet] 2018;562(7725):128–32. Available from: https://doi.org/10.1038/s41586-018-0522-3.
[57] Rafii S, Butler JM, Ding B-S. Angiocrine functions of organ-specific endothelial cells. Nature [Internet] 2016;529:316. Available from: https://doi.org/10.1038/nature17040.
[58] Calabrese C, Poppleton H, Kocak M, Hogg TL, Fuller C, Hamner B, et al. A perivascular niche for brain tumor stem cells. Cancer Cell [Internet] 2007;11(1):69–82. Available from: <http://www.sciencedirect.com/science/article/pii/S1535610806003692>.
[59] Wada T, Haigh JJ, Ema M, Hitoshi S, Chaddah R, Rossant J, et al. Vascular endothelial growth factor directly inhibits primitive neural stem cell survival but promotes definitive neural stem cell survival. J Neurosci [Internet] 2006;26(25):6803–12. Available from: <http://www.jneurosci.org/content/26/25/6803>.
[60] Zhu TS, Costello MA, Talsma CE, Flack CG, Crowley JG, Hamm LL, et al. Endothelial cells create a stem cell niche in glioblastoma by providing NOTCH ligands that nurture self-renewal of cancer stem-like cells. Cancer Res [Internet]

2011;71(18):6061–72. Available from: <http://cancerres.aacrjournals.org/content/71/18/6061>.

[61] Lu J, Ye X, Fan F, Xia L, Bhattacharya R, Bellister S, et al. Endothelial cells promote the colorectal cancer stem cell phenotype through a soluble form of Jagged-1. Cancer Cell [Internet] 2013;23(2):171–85. Available from: <http://www.sciencedirect.com/science/article/pii/S1535610813000317>.

[62] Hanahan D, Weinberg RA. Hallmarks of cancer: the next generation. Cell [Internet] 2011;144(5):646–74. Available from: <http://www.sciencedirect.com/science/article/pii/S0092867411001279>.

[63] Gao H, Chakraborty G, Zhang Z, Akalay I, Gadiya M, Gao Y, et al. Multi-organ site metastatic reactivation mediated by non-canonical discoidin domain receptor 1 signaling. Cell [Internet] 2016;166(1):47–62. Available from: <http://www.sciencedirect.com/science/article/pii/S0092867416307383>.

[64] Cao Z, Scandura JM, Inghirami GG, Shido K, Ding B-S, Rafii S. Molecular checkpoint decisions made by subverted vascular niche transform indolent tumor cells into chemoresistant cancer stem cells. Cancer Cell [Internet] 2017;31(1):110–26. Available from: <http://www.sciencedirect.com/science/article/pii/S1535610816305530>.

[65] Pan H, Gray R, Braybrooke J, Davies C, Taylor C, McGale P, et al. 20-year risks of breast-cancer recurrence after stopping endocrine therapy at 5 years. N Engl J Med 2017;377(19):1836–46.

[66] Goss PE, Chambers AF. Does tumour dormancy offer a therapeutic target? Nat Rev Cancer 2010;10(12):871.

[67] Shibue T, Brooks MW, Weinberg RA. An integrin-linked machinery of cytoskeletal regulation that enables experimental tumor initiation and metastatic colonization. Cancer Cell 2013;24(4):481–98.

[68] Yao H, Price TT, Cantelli G, Ngo B, Warner MJ, Olivere L, et al. Leukaemia hijacks a neural mechanism to invade the central nervous system. Nature [Internet] 2018;560(7716):55–60. Available from: https://doi.org/10.1038/s41586-018-0342-5.

[69] Wiencken-Barger AE, Mavity-Hudson J, Bartsch U, Schachner M, Casagrande VA. The role of L1 in axon pathfinding and fasciculation. Cereb Cortex 2004;14(2):121–31.

[70] Schröder C, Schumacher U, Fogel M, Feuerhake F, Müller V, Wirtz RM, et al. Expression and prognostic value of L1-CAM in breast cancer. Oncol Rep 2009;22(5):1109–17.

[71] Cuddapah VA, Robel S, Watkins S, Sontheimer H. A neurocentric perspective on glioma invasion. Nat Rev Neurosci 2014;15(7):455.

[72] Cortina C, Palomo-Ponce S, Iglesias M, Fernández-Masip JL, Vivancos A, Whissell G, et al. EphB–ephrin-B interactions suppress colorectal cancer progression by compartmentalizing tumor cells. Nat Genet [Internet] 2007;39:1376. Available from: https://doi.org/10.1038/ng.2007.11.

[73] Watkins S, Robel S, Kimbrough IF, Robert SM, Ellis-Davies G, Sontheimer H. Disruption of astrocyte–vascular coupling and the blood–brain barrier by invading glioma cells. Nat Commun 2014;5:4196.

[74] Lawler J. Thrombospondin-1 as an endogenous inhibitor of angiogenesis and tumor growth. J Cell Mol Med 2002;6(1):1–12.

[75] Weinstat-Saslow DL, Zabrenetzky VS, VanHoutte K, Frazier WA, Roberts DD, Steeg PS. Transfection of thrombospondin 1 complementary DNA into a human breast carcinoma cell line reduces primary tumor growth, metastatic potential, and angiogenesis. Cancer Res 1994;54(24):6504–11.

[76] Massagué J. TGFβ in cancer. Cell 2008;134(2):215–30.

[77] Carbonell WS, DeLay M, Jahangiri A, Park CC, Aghi MK. β1 integrin targeting potentiates antiangiogenic therapy and inhibits the growth of bevacizumab-resistant glioblastoma. Cancer Res [Internet] 2013;73(10):3145–54. Available from: <http://cancerres.aacrjournals.org/content/73/10/3145>.

CHAPTER

4

Pericyte mimicry: an embryogenesis-derived program of extravascular tumor cell migration

Claire Lugassy[1], Hynda Kleinman[2] and Raymond Barnhill[1]

[1]Department of Translational Research, Curie Institute, Paris, France, [2]Department of Molecular Medicine and Biochemistry, The George Washington School of Medicine, Washington, DC, United States

Normal science ... is predicated on the assumption that the scientific community knows what the world is like. Normal science often suppresses fundamental novelties because they are necessarily subversive of its basic commitment [1].

Introduction

Interactions between tumor cells and blood vessels

Cancer growth refers to uncontrolled tumor cell proliferation, while cancer metastasis refers to the migration of cancer cells out of the primary (initial tumor) lesion to another part of the body where new tumor growth occurs [2]. Primary tumors and metastatic tumors may comprise various different populations of tumor cells and thus represent distinct

Tumor Vascularization
DOI: https://doi.org/10.1016/B978-0-12-819494-2.00004-3

pathologies with different therapeutic approaches. These findings suggest that the cells that migrate out of the primary tumor are distinct and generally more aggressive. Cancer cells can use different migratory strategies depending on varying environments to exit the primary tumor mass, invade surrounding tissues, and metastasize to distant sites. However, the precise mechanism of tumor growth and tumor cell migration (or invasion), and in particular the role of the tumor vasculature during these two fundamental steps in cancer progression, remains unclear [3].

There was considerable debate about the potential mechanisms of cancer metastasis until the end of the 19th century. The mechanism of intravascular dissemination of cancer was finally recognized and is still considered as the only accepted paradigm explaining metastasis, and constitutes the basis for "normal science" [1] in cancer research. Furthermore, in 1971, Judah Folkman suggested the paradigm of angiogenic tumors, demonstrating that without angiogenesis a tumor cannot grow [4]. Folkman indicated also that tumor vascularity correlated with metastases [5]. This was the opening of a new and important cancer research area that was widely accepted as "normal science." Therefore the relationships between a tumor and its vasculature were mainly focused on (1) angiogenesis for tumor growth at both primary and secondary sites and (2) intravascular circulation of tumor cells toward metastatic sites.

However, for more than 20 years, several of the above assumptions from the present "normal science" have been questioned:

It was shown in 1994 that tumor vascularity was not correlated with melanoma metastasis [6], questioning the fact that angiogenesis is a prognostic marker. In addition, while it was assumed that tumor cells in contact with vessel lumina were in the process of intravasation, it was shown in the late 1990s that such a vessel–tumor interaction could have a completely different biological interpretation. Indeed, from independent research laboratories, the presence of tumor cells along the abluminal vascular surface has defined two new areas of cancer research: (1) vascular co-option and (2) angiotropism, pericyte mimicry (PM), and extravascular migratory metastasis (EVMM).

These two closely related areas have raised important questions about different aspects of cancer biology, which may be complementary during tumor progression. The field of vascular co-option, defined by Pezzella and colleagues [7] and described in depth in another chapter, has revealed the fundamental importance of nonangiogenic tumors, questioning the long-standing paradigm of Folkman that all tumor growth is dependent on angiogenesis [4]. Therefore the proposed pathobiological significance of vascular co-option is that tumors utilize alternative mechanisms besides angiogenesis to obtain nutrients for growth, via local tumor invasion and proliferation along co-opted vessels.

On the other hand, research on angiotropism, PM, and EVMM, opened by Lugassy and Barnhill [8], has questioned the assumption that all tumor cells exclusively use intravascular (blood or lymphatic vessels) dissemination for their spreading and formation of metastasis. The main proposed pathobiological significance of this area of research is that tumor cells use embryogenesis-derived mechanisms to spread via progressive and continuous migration along the abluminal vascular surfaces (and via other extravascular pathways) to nearby or more distant sites, without entering into the vascular channels.

History

In 1829 the French physician Joseph Récamier was the first to use the term metastasis and referred to the spread of tumor cells along the external surfaces of vascular channels rather than within them [9]. In 1907 Handley cited Borst who had noted "the tendency of melanotic sarcoma to spread along the perivascular tissues immediately outside the blood vessels" [10]. In 1996 Lugassy et al. reported direct contact between tumor cells and endothelium (angiotumoral complex) via a lamina-rich in laminin. They also noted the pericytic location of tumor cells, which "may promote the migration of tumor cells in contact with vessels" [11]. Since 2002, Lugassy and Barnhill have shown that angiotropism (the pericytic-like location of tumor cells) is a microscopic marker of migration along the abluminal vascular surface [8]. In 1999 these authors termed the tumor cell dissemination via different pathways outside the vessel lumina as Extravascular Migratory Metastasis or EVMM [12]. In 2013 they introduced the term "pericytic mimicry" (PM) in order to describe more specifically the embryogenesis-derived biological process of tumor spread along the periendothelial, abluminal vascular surface, competing with, replacing, and mimicking pericytes [13].

Such findings have recently been confirmed by several groups [14–16].

Angiotropism, PM and EVMM

PM is a mechanism of progressive tumor migration (or tumor spread) along the abluminal surface of vessels without intravasation. During PM, tumor cells have an anatomical location along the external surfaces of vessels defined as angiotropism, which is observed in histopathological images. Angiotropism is a marker of PM in several experimental models. PM may result in local or long-distance migration inside or outside the primary and secondary tumors. PM is a part of EVMM which can also, depending on the microenvironment, occur via other tissue tracks, such as nerves.

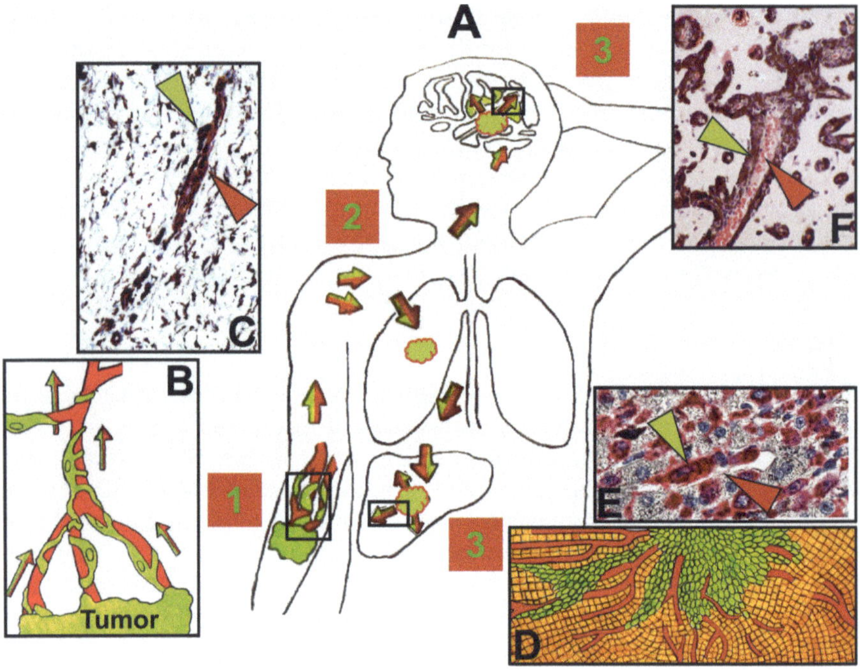

FIGURE 4.1 Diagram of PM and EVMM.
(A) Overview diagram of human body. (B) Diagram of PM: tumor cells migrate in a pericytic location along the abluminal vascular surface at the advancing front of the tumor. (C) Corresponding human histopathological image of angiotropism/PM at the advancing front of the primary melanoma. Double immunostaining of melanoma cells by melan-A (red chromogen) and vascular channels by CD31 (brown-black chromogen). (D) Diagram of the replacement histopathological growth pattern in liver metastasis. Note at the advancing front of the metastasis the progression of green tumor cells replacing hepatocytes (*orange*) along red vessels. (E) Double immunostaining of a corresponding human histopathological image at the advancing front of a liver metastasis. Melanoma cells express HMB45 along endothelial cell expressing CD31 (DAB brown chromogen). (F) Human melanoma brain metastasis. Striking angiotropism of pigmented melanoma cells in a pericytic location along microvessels containing erythrocytes.
Green arrowheads indicate tumor cells, *red arrowheads* indicated vessels. *Bicolor arrows* indicate a schematic view of PC. (*1*) PM at the advancing front of the primary tumor. (2) EVMM via PM throughout the body. (3) Metastases sites.

Angiotropism, PM, and EVMM are three complementary terms referring to a mechanism of progressive and continuous tumor cell migration without entering the vascular lumen (Fig. 4.1).

- Angiotropism

Angiotropism is defined histologically as tumor cells closely associated with the endothelium of vessels in a pericytic location without intravasation, detected at the advancing front of the tumor [8]. Angiotropism is observed in both primary and secondary melanoma.

- PM

Pericytes are modified smooth muscle cells that wrap closely around small blood vessels, regulating and supporting the microvasculature through direct endothelial contact. Angiotropism involves PM, the spreading of tumor cells along the abluminal vascular surface [8]. Angiotropic tumor cells localize along the abluminal vascular surfaces by competition with, replacement of, and mimicking of pericytes in a number of ways: (1) tumor cells are recruited along the abluminal vascular surface instead of pericytes and (2) tumor cells migrate along the abluminal surface (Fig. 4.1).

Tumor cells demonstrate PM in in vitro and in vivo models [8]. However, tumor cells never stabilize neovessels nor the vascular basement membrane matrix as do pericytes. PM is an essential form of EVMM, allowing tumor cells to progress along vessels with all the needed nutrients and oxygen, and such a process avoids the destruction of tumor cells observed in hematogenous dissemination [17].

Notably, pericytes are being increasingly studied for their role in tumor formation, growth, invasion, and metastasis [18]. However, the respective role of pericytes and angiotropic tumor cells in tumor progression requires more investigation.

- EVMM

The continuous migration of tumor cells toward secondary sites (or between secondary sites), without entering inside the lumina of vascular channels, represents EVMM. EVMM is an important alternative mechanism or possibly the primary mechanism of progressive, "step-by-step" melanoma migration (Fig. 4.1) versus the "express" intravascular tumor cell dissemination [8,17]. In addition to PM along vessels, EVMM can also occur via other tissue tracks [8].

PM in melanoma

Histopathology

Angiotropism is the histopathological marker of PM. Angiotropism is characteristically recognized at the advancing front of a tumor, and has been observed in cutaneous, uveal, and conjunctival melanoma. Angiotropism is present in primary melanoma, satellites and in transit metastasis, lymph nodes and distant metastases [8,19–25].

Histological criteria

Angiotropism in cutaneous melanoma is defined as follows: (1) tumor cells closely aligned along the external (abluminal) surfaces of the endothelium of microvascular and/or lymphatic channels, either in linear single-cell or multilayered cellular aggregates; this is also characteristically a "pericytic" location; (2) the latter occurring at the tumor–stromal interface (the advancing front) of a tumor or in the nearby tissue (usually at least 0.5–2 mm away from the tumor); and (3) no evidence of intravascular or intralymphatic (intravasation of) tumor aggregates. Angiotropism is recognized at the advancing front of the tumor or in nearby tissue so that it is clearly distinguished from entrapment or engulfment of vessels by the tumor [8,19]. Although the latter feature may in fact constitute angiotropism in some instances, there is no means at present to verify if such entrapment constitutes angiotropism biologically or not (Figs. 4.2 and 4.3).

In primary uveal melanoma, angiotropism is characterized by migration of melanoma cells from the primary intraocular uveal site along the

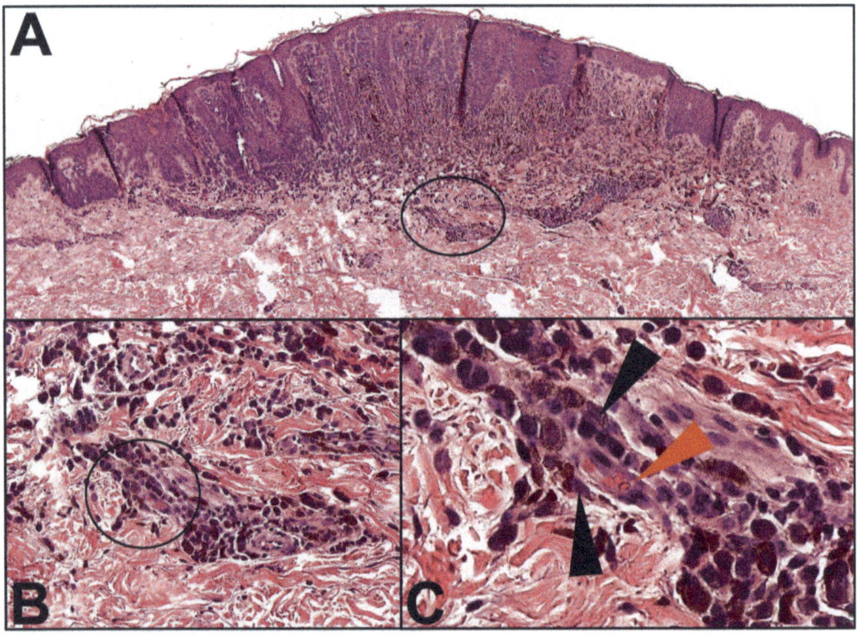

FIGURE 4.2 Angiotropism in primary cutaneous melanoma.
(A) Low magnification of the tumor. Encircled area: angiotropism at the advancing front (the base) of the lesion. (B and C) Higher magnification of the advancing front. Note the angiotropic melanoma cells (*black arrows*) associated with vascular channels (*red arrow*).

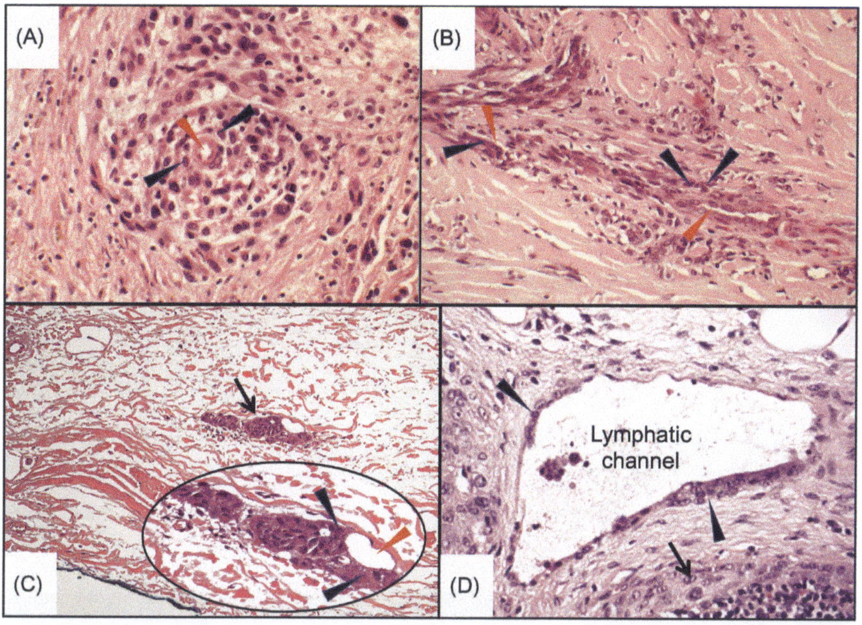

FIGURE 4.3 Angiotropism in primary and loco-regional melanoma.
(A and B) Melanoma cells at the advancing front of their respective primary melanoma. Melanoma cells aligned around (A) and along (B) the abluminal surface of microvascular channels in a pericytic location. (C) Angiotropic microscopic satellite some distance from the primary melanoma (about 1 mm). The microsatellite (*black arrow*) is separated from the primary melanoma by uninvolved dermal collagen. *Inset*: At higher magnification, melanoma cells (*dark purple cells*) are cuffing the abluminal surface of the microvascular channel. This small tumor aggregate of melanoma cells shows angiotropism and is separated from the primary melanoma by uninvolved dermal collagen. (D) Metastatic melanoma in a sentinel lymph node from primary cutaneous melanoma. A large lymphatic channel in the fibrous capsule of the lymph node shows striking lymphangiotropism. Melanoma cells are disposed along the external surfaces of endothelial cells lining the lymphatic channel, both as single cells and in aggregates (*black arrowheads*). Degenerative cellular debris is noted within this channel. In addition, a crescent-shaped deposit of melanoma cells is present in the peripheral sinus of the lymph node in the lowermost portion of this field (*black arrow*).
Black arrowheads: melanoma cells. *Red arrowheads*: vascular channels. *Source (D) Courtesy Dr. A. Cochran UCLA.*

abluminal surfaces of vascular channels traversing the sclera (including the endothelial-lined channels of Schlemm) [20] (Fig. 4.4).

In common with cutaneous melanoma, conjunctival melanoma may show angiotropism at the advancing front of the primary melanoma and loco-regional satellite-in transit metastases in nearby conjunctiva or eyelid skin [21] (Fig. 4.5).

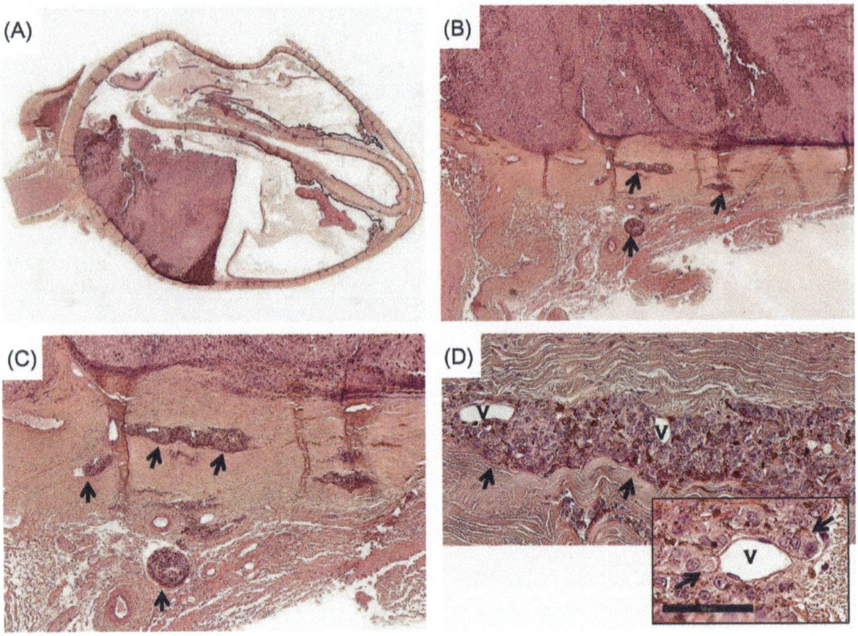

FIGURE 4.4 Angiotropism in primary uveal melanoma. (A) Large nodular pigmented tumor involving the posterior uveal region. (B) Angiotropism of melanoma (*arrow*) along posterior ciliary vessels within the sclera at scanning magnification. On the outer surface of the sclera, a larger aggregate of angiotropic melanoma cells constitutes extraocular extension and a satellite-in transit metastasis. (C) Melanoma cells are disposed along the abluminal surfaces of the vascular channels (arrows) without intravasation. (D) High magnification shows multilayered arrangements of pigmented epithelioid melanoma cells (*arrows*) surrounding this ciliary vascular channel. V = intravascular lumina. *Inset*: Abluminal melanoma cells (*arrows*) are circumferentially and directly positioned along and in contact with the external surface of this endothelial-lined vascular channel (V). These melanoma cells are in contact with the basement membrane of the endothelial cells. *Source: A–D from Ref. [20].*

Melanoma liver metastases

Liver metastases from colorectal cancers [7] and melanoma, both uveal [22] and cutaneous (manuscript in preparation), show distinctive "histopathological growth patterns" (HGPs): "replacement," "desmoplastic," "pushing," and two rare variants—the "sinusoidal" and "portal" HGPs [7]. In particular, the replacement HGP is defined by tumor cells that progressively infiltrate and replace hepatocytes at the tumor–stromal interface (the advancing front of the metastasis) in a "pericytic" location along the abluminal surfaces of sinusoidal vessels or venules of the portal tracts [22]. The replacement pattern represents approximately 50% of the liver metastases from cutaneous melanoma

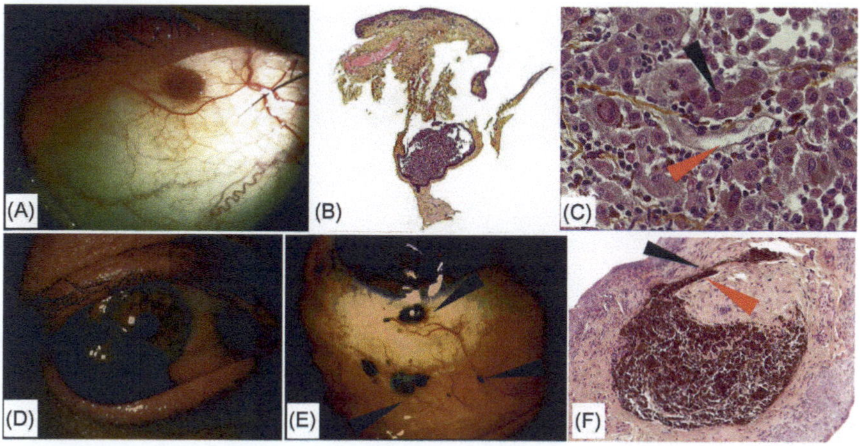

FIGURE 4.5 Angiotropism in conjunctival melanoma. (A) Satellite-in transit metastasis involving nasal bulbar conjunctiva of the left eye. (B) 1.3 mm in diameter micronodular lesion (*back arrow*) is in the deep substantia propria of the conjunctiva some distance away from the site of a previously excised primary conjunctival melanoma. (C) Melanoma cells (*black arrowhead*) are external to vascular channels (*red arrowhead*) indicating PM. (D) Primary bulbar conjunctival melanoma with a bulky pigmented nodular tumor involving the temporal bulbar conjunctiva. (E) Satellite-in transit metastases from this bulbar conjunctiva (D). Some of the lesions (*black arrowheads*) are associated with microvascular channels (*red arrowheads*). (F) Histological examination shows a well-circumscribed subepithelial nodular aggregate of heavily pigmented melanoma cells. Note angiotropism of melanoma cells (*black arrowheads*) in a pericytic location along a microvascular channel (*red arrowheads*), indicating PM. *Source: A–F from Ref. [21].*

(manuscript in preparation), and 75% of metastases associated with uveal melanoma [22] (Figs. 4.1 and 4.6). Importantly, the replacement HGP in liver metastases is associated with a poorer prognostic [7,22].

Experimental models of PM

Several experimental studies have shown that melanoma angiotropism is a microscopic marker indicative of migration (spread) along the abluminal vascular surfaces, that is, PM.

- In vitro studies

  3D cocultures of melanoma cells with endothelial tubules

In order to study the relationships between human melanoma cells and vascular endothelium, melanoma cells were cultured with endothelial cells that had already formed endothelial tubules on a tumor basement membrane matrix (Matrigel) [8,26]. Real time imaging

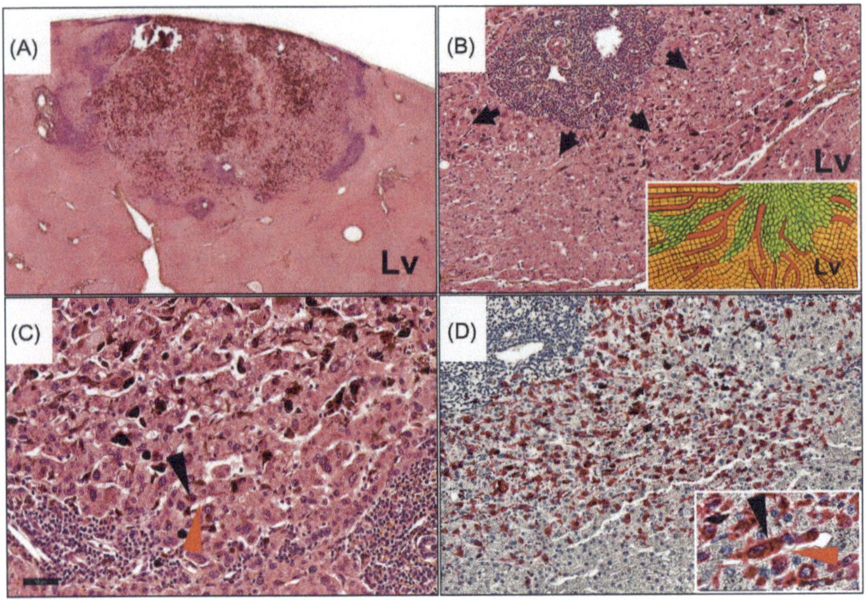

FIGURE 4.6 Replacement histopathological growth pattern in a liver metastasis from uveal melanoma.
(A) Scanning magnification shows a discrete metastasis abutting the liver capsule. Lv, surrounding uninvolved liver parenchyma. (B) The interface (broad advancing front extending from upper right quadrant to lower left quadrant) between the metastasis (*upper left quadrant*) and surrounding liver parenchyma (Lv) (*lower right quadrant*) is poorly defined. Melanoma cells (detected by the presence of cytoplasmic melanin and cytological atypia) are dispersed throughout this peripheral interface as single cells (*black arrows*) and small clusters beyond the main portion of the metastasis. These cells replace hepatocytes in the hepatic plates without altering their architecture. (C) Melanoma cells extending into the surrounding hepatic plates some distance from the metastasis along the sinusoidal vessels in a "pericytic" location, indicating PM. *Red arrowhead* indicates vascular lumen. *Black arrowhead* identifies melanoma cells. *Inset*: Diagram of the replacement histopathological growth pattern in liver metastasis. Note at the advancing front of the metastasis the progression of green tumor cells replacing hepatocytes (*orange*) along red vessels. (D) HMB45 immunostain with red chromogen highlights melanoma cells extending into the surrounding liver parenchyma. *Inset*: Melanoma cells expressing HMB45 are disposed along the abluminal endothelial surface of a sinusoidal vessel in the hepatic parenchyma (*black arrowhead*). The endothelial cell lining is highlighted by CD31 (DAB brown chromogen, *red arrowhead*). *Source: A–D from Ref. [22].*

demonstrated angiotropism of melanoma cells toward the vascular tubules and migration along the abluminal vascular surfaces, that is, PM within 23–24 hours [8] (Fig. 4.7A and B). Similar results have been found with cutaneous and uveal melanoma cells [27]. This model has been used with cancer cells from other malignancies suggesting that

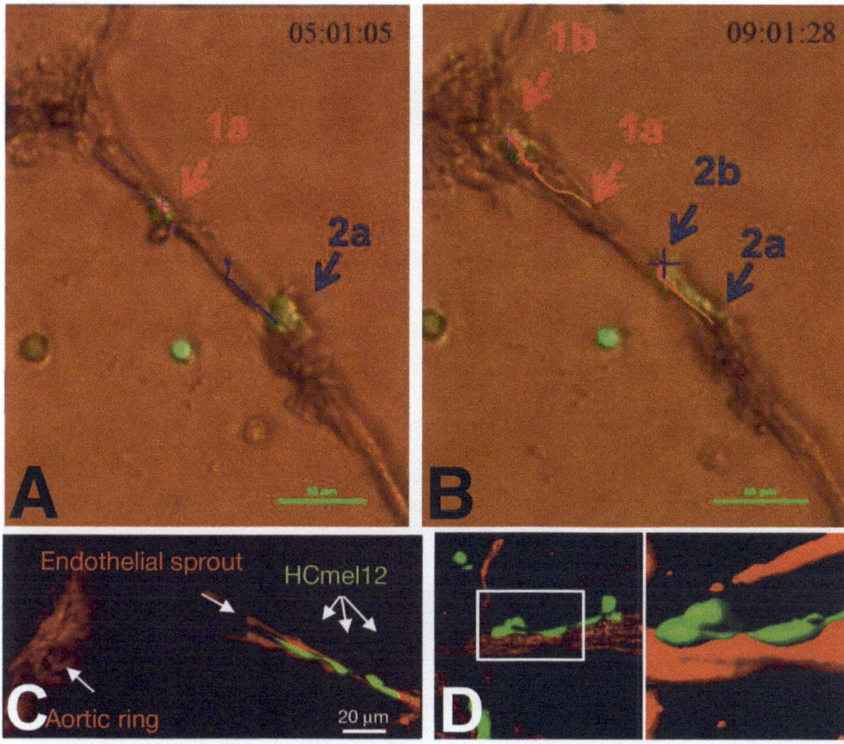

FIGURE 4.7 In vitro and ex vivo studies of PM. (A and B) 3D cocultures of melanoma cells with endothelial tubules. Single-cell analysis of GFP (*green*) melanoma cells engaging in PM along vascular tubules. Time-lapse images of melanoma cells 1 and 2 during 4 h of PM. Melanoma cells 1 and 2 have moved along the endothelial tubule from points 1a to 1b. Note the mesenchymal shape of the melanoma cells in (B). Scale bars = 60.
(C and D) Melanoma cells and rat aortic rings in 3D culture. GFP (*green*)-labeled melanoma cells in a pericyte-like association with tomato lectin (*red*)-labeled endothelial cells sprouting from aortic rings. (D) On the right, high magnification of the inset. *Source: A–B from Ref. [8]; C–D from Ref. [23].*

other cell types beyond melanoma may use a similar mechanism to form metastatic lesions [14,28].

- Ex vivo studies

  Melanoma cells and rat aortic rings in 3D culture

Aortic rings were cocultured in 3D in Matrigel with human melanoma cell aggregates. After 24–48 hours, human melanoma cells demonstrated angiotropism and spreading along endothelial sprouts from the aortic ring explants, that is, angiotropism and PM along neovessels as seen in Fig. 4.7C and D [8,23].

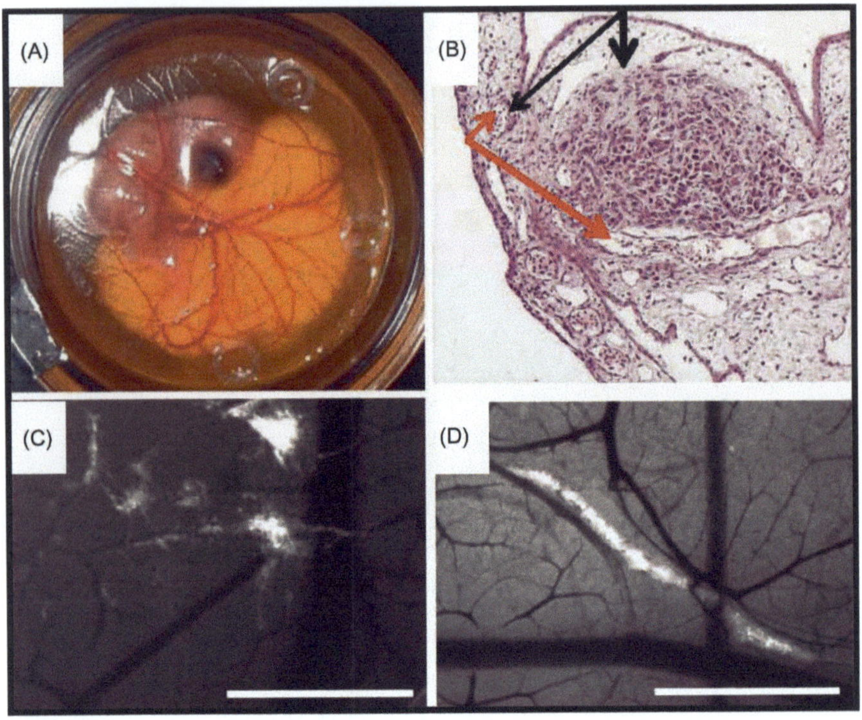

FIGURE 4.8 The shell-less chick CAM assay. (A) The chicken eggs into a glass dish with GFP melanoma cells applied to the surface of the CAM. Melanoma cells spreading along (B) and jumping between (C) microvascular channels, indicating PM. Bars = 1 mm. (D) Histopathology confirmed angiotropism of melanoma cells (*black arrows*) around microvessels (*red arrows*) without intravasation. Note one tumor cell showing angiotropism (*thin black arrow*). *Source: A–D from reference 28.*

Melanoma cells on murine ear explants

An ear tissue invasion assay was established in which fluorescent green melanoma cells were seeded onto murine ear tissue explants and allowed to migrate into the dermis. Confocal laser-scanning immunofluorescence microscopy demonstrated the migration of melanoma cells into the dermis along vessels, demonstrating PM [23].

The shell-less chick chorioallantoic membrane (CAM) model

A 4-day-old fertilized egg is cracked into a glass dish, and at day 10 melanoma cells are applied to the surface of the CAM. PM of melanoma cells along CAM vessels demonstrates melanoma angiotropism similar to that observed in human angiotropic melanoma observed in biopsies (Fig. 4.8). This finding confirms PM along vessels without intravasation. Several studies have also suggested that tumor cells are able to migrate from one vessel to another, perhaps in order to maintain a particular direction [8,29].

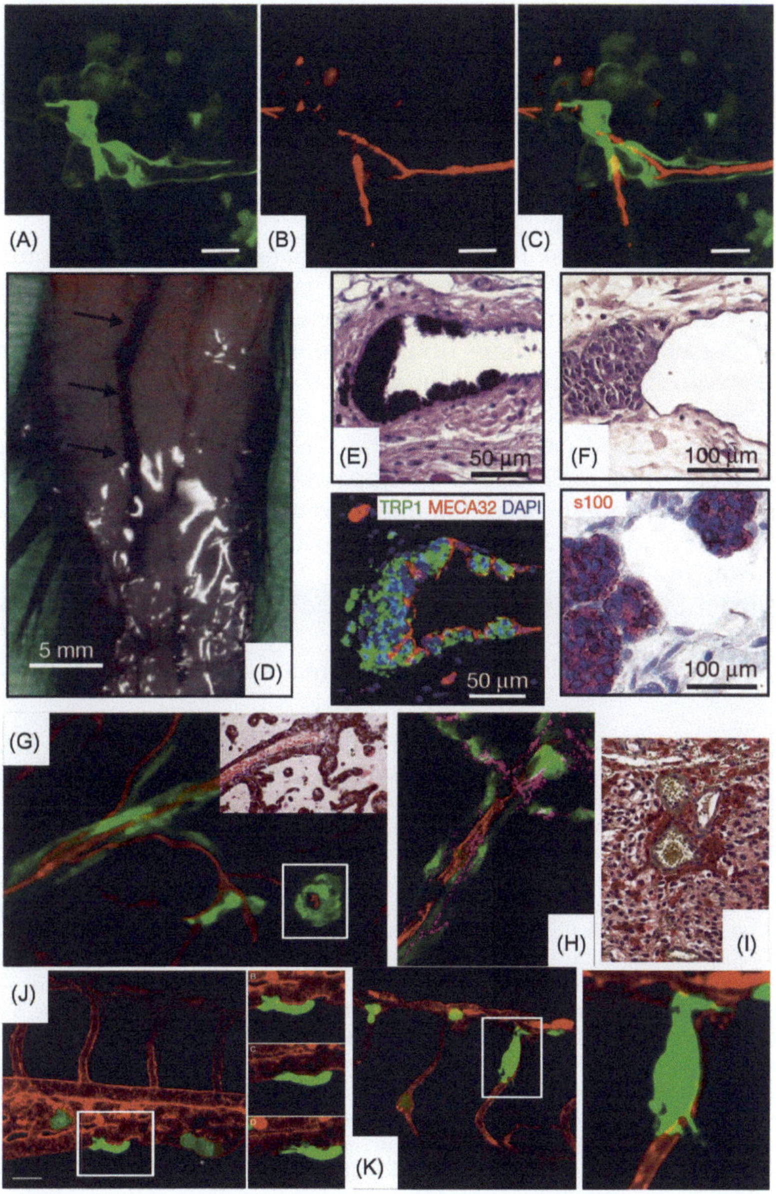

FIGURE 4.9 In vivo studies of PM.
(A–C) Intravital imaging of PM in the mouse ear skin. Two-color intravital imaging of GFP melanoma tumors in the mouse ear by 2PE microscopy. (A) Green melanoma tumor cells. (B) Red (tomato lectin) dermal vessels. (C) Overlay image showing PM of individual melanoma cells spreading along vessels [10]. Scale bars, 50 µm. (D–F) PM and angiotropism in a genetically engineered mouse model. (D) Macroscopically visible melanoma cell expansion along 25 mm of a dermal blood vessels (*arrows*). Histologic and

- In vivo models (Fig. 4.9)

Intravital imaging of PM in the mouse ear skin

After intradermal injection of human melanoma cells in the mouse ear skin, two-color intravital imaging allows clear observation of angiotropism and PM without intravasation of individual melanoma cells spreading along vessels [30] (Fig. 4.9A–C).

Murine cutaneous melanoma model

UV irradiation of skin with DMBA-initiated primary melanomas in HGF-CDK4 genetically engineered mice promoted expansion of melanoma angiotropism in the primary tumor and PM along abluminal blood vessel surfaces. These findings demonstrated the association of these processes with increased lung metastases [23]. This model allowed the observation of macroscopically visible melanoma cells spreading along dermal blood vessels, that is, PM. This was confirmed by histopathology showing significant angiotropism analogous to human angiotropism. Three months after the initiation of the experiment, melanoma cells had spread over 25 mm along a dermal blood vessel (Fig. 4.9D–F).

Murine brain melanoma model

Human melanoma cells were injected directly into the murine brain as a primary site of tumor development. The interaction of green fluorescent melanoma cells with red fluorescent vessels was analyzed over

◀ immunohistochemical analysis of angiotropism in the murine model (C) and in a human primary melanoma (D). Note the similar angiotropic images in (C) and (D). (G–I) Human melanoma cells in a murine melanoma model brain model. (G) Montage juxtaposing PM in the murine brain melanoma model and a human sample of a melanoma metastatic to the brain. On the left part of the image, in the murine model at 4 weeks, PM of green GFP-labeled melanoma cells are visible along red vessels (*red tomato lectin*). Juxtaposed to the right of the same image: human sample of a melanoma metastatic to the brain. Microvessels (filled by red blood cells) are extensively coated by a layer of melanin–containing melanoma cells, without intravasation, demonstrating significant angiotropism and PM. Note the correspondence of the two juxtaposed images. Scale bars: 100 μm. *Inset*: XZ cross section of the murine brain vessel confirming that melanoma cells are external to the vessels, without intravasation. (H) Expression of Serpin B2 in the murine brain melanoma model. Strong colocalization of Serpin B2 (*purple*) to the angiotropic GFP melanoma cells (*green*) along vessels (*red tomato lectin*). Scale bars, 50 μm. (I) Expression of Serpin B2 by angiotropic melanoma cells in a human brain melanoma metastasis. Striking expression of SERPIN B2 (*dark red*) by melanoma cells is observed predominately along the abluminal surfaces of microvascular channels. (J and K) Intravital observation of PM of melanomas in a zebrafish xenograft. (J) Xenograft with GFP green human cutaneous melanoma cells. Angiotropism/PM of green GFP tumor cells along the external surface of the caudal vein (*red tomato lectin*). On the right: time-lapse images of the angiotropic cell in white square taken at 0, 4, and 8 h after the beginning of the imaging. (K) Xenograft with GFP human uveal melanoma cells. Similar image along the external surface of red intersegmental vessels. Note single (J and K) and collective (K, *white square*) tumor cell migration along vessels. *Sources: A–C from reference 30; D–F from reference 23; G–I from reference 31; J–K from reference 27.*

time. 3D images revealed progressive and extensive spread of melanoma cells along the abluminal surfaces of the microvascular channels without intravasation, that is, PM [31]. This study also documented the perivascular expression of Serpin B2 by angiotropic melanoma cells in the murine brain and in human melanoma brain metastases from patients with primary cutaneous melanoma (Fig. 4.9G–I).

PM of human cutaneous and uveal melanomas in a zebrafish xenograft

At 2 dpf, tumor cells were injected into the yolk sac of zebrafish larvae. At 30 hours postinjection, intravital imaging indicated that cutaneous and uveal melanoma cells spread similarly along the abluminal vascular surfaces, exhibiting changes in the structural physiology suggesting epithelial–mesenchymal transition (EMT) [27]. Both single and collective cell migration were observed (Fig. 4.9J and K).

Prognostic significance of angiotropism and PM

Angiotropism, the histopathological marker of PM showing the proximity of the tumor cell to the abluminal vessel wall, is an independent prognostic factor predicting risk for metastasis and survival in cutaneous melanoma. This has been shown for micro- or macroscopic satellites, in transit metastases, lymph node metastases, and distant metastases [8,19,23,24,32]. Angiotropism is also a comparable prognostic factor in uveal melanoma [20] and is predictive of the development of metastases and overall survival. In addition, the replacement HGP in liver metastases, that is, PM, from both cutaneous and uveal metastases, is predictive of significantly diminished survival on both univariate and multivariate analysis [22] (manuscript in preparation).

Molecular findings

- Gene expression profiling of human angiotropic primary melanoma

Using a previously constructed melanoma gene expression microarray, 15 genes potentially critical to PM have been identified [8,33]. These 15 genes were classified according to their relationship to (1) neural crest cell migration (TCOF1, NEIL3, AHNAK, KCTD11, HMMR, CEBPA, AQP3), (2) cell migration of other malignant tumors with neural crest origin (ECT2, AGAP2, GLS), (3) cell motility and/or migration (FGD3, F10, DBF4, AHNAK FNBP1L), and (4) neurotropism (KIF14).

- Gene expression triggered by PM in vitro

Microarray analysis on the vitro coculture model (melanoma cells cultured with already formed endothelial tubules) of PM revealed 28

TABLE 4.1 Gene expression of angiotropic melanoma cells interacting with the abluminal vascular surface in a 3D coculture model.

Cell migration	CCL2, ICAM1, IL6, RGNEF, RANBP9, PDGFB
Cancer progression	CCL2, ICAM1, SELE, TRAF1, IL6, SERPINB2, CXCL6, BLID, MALT1, UPF1, PLAA, RGNEF, ZXDC
EMT	CCL2, IL6, ICAM1, PDGFB
Embryonic/(cancer) stem cell properties	CCL2, PDGFB, EVX1, CFDP1, RANBP9
Pericvtic recruitment	PDGFB
Inflammation	CCL2, IL6, TRAF1, CXCL6, SELE, ICAM1, SERPINB2, SLC7A2, C2CD4B, PDGFB

Subclassification of 20 from the 28 differentially expressed genes in six categories.

differentially expressed genes [30]. Among them, 20 genes demonstrated properties linked to (1) cell migration (CCL2, ICAM1; IL6, RGNEF, RANBP9, and PDGFB), (2) cancer progression and metastasis (CCL2, ICAM1, SELE, TRAF1, IL6, SERPINB2, CXCL6, BLID, MALT1, UPF1, PLAA, RGNEF, ZXDC), (3) EMT (CCL2, IL6, ICAM1, PDGFB), (4) embryonic and/or cancer stem cell (CSC) properties (CCL2, PDGFB, EVX1, CFDP1, and RANBP9), (5) pericyte recruitment (PDGFB), and (6) inflammation (CCL2, IL6, TRAF1, CXCL6, SELE, ICAM1, SERPINB2, SLC7A2, C2CD4B, PDGFB) (Table 4.1) [8,34]. Notably, SERPINB2, a gene promoting cancer cell survival and vascular co-option in a murine model of brain metastasis from lung and breast cancer [35], was also associated with PM in the murine brain melanoma model described above [31].

- PDGF-β/PDGFR-β

The ligand/receptor pair platelet-derived growth factor β (PDGF-β/PDGFR-β) is essential for pericyte recruitment [18,36], and PDGF-β is upregulated during the coculture of endothelial cells with melanoma cells [8,34]. Pericytes express PDGFR-β, and in normal angiogenesis, the recruitment of pericytes stabilizes the neovessels [36]. Melanoma cells express this same receptor, and PDGFR-β is expressed by angiotropic melanoma cells in human melanoma samples.

- Pericyte-marker expression by melanoma cells

Other markers of pericytes, CD146 and NG2, are also be expressed by melanoma cells [34]. Finally, flow cytometry demonstrated the expression of pericyte markers (CD44, CD73, CD105, and CD144) by the C8161 melanoma cell line [34].

- Role of neutrophilic inflammation on angiotropism and PM in melanoma

Experimental evidence that neutrophilic inflammation promotes angiotropism, PM, and metastatic spread came from a genetically engineered melanoma model [23]. UV irradiation promotes PM and metastatic dissemination through TLR4/MyD88-driven neutrophilic inflammation initiated by HMGB1 release from UV-damaged keratinocytes.

Angiotropism, PM, and EVMM in non-melanoma tumors

Many other tumor types besides melanoma exhibit angiotropism, PM, and EVMM as well as migration along other types of tissue tracts. For example, glial cells, like melanocytes, derive from the neural crest cell. Invading glioma cells are known to follow distinct anatomic structures within the central nervous system, including the abluminal surface of blood vessels [37,38], exhibiting the same phenotypic PM as angiotropic melanoma cells [28,38,39]. Glioma cells can migrate considerable distances without employing intravascular dissemination [37,39], that is, EVMM. The incidence of metastases is low, and this can be due to the short survival of involved patients given the local aggressiveness of brain tumors [40].

Very little is understood about pathways that regulate the blood–brain barrier (BBB) permeability in the normal brain or in brain tumors [41]. PM, without intravasation and therefore without needing to breach the BBB, could represent an explanation for brain metastases from multiple tumors. Notably, leukemia cells in the circulation cannot breach the BBB and instead they reach the central nervous system by migrating along the external surface of blood vessels, that is, PM [15] (Fig. 4.10C).

In multiple primary tumors and metastases, some HGPs of tumor invasion are similar to PM, involving mainly tumor migration along vessels [7]. For example, the lepidic pattern in the lung and the replacement pattern in liver metastases from colon cancer, breast cancer [7], and from uveal and cutaneous liver metastases all demonstrate examples of PM [22] (manuscript in preparation). PM has also been shown as a mechanism for initiating metastatic colonization in multiple organs by metastatic dormant cells after they exit a latency period [14].

Several types of malignant cells, such as prostate and pancreatic cancers, spread along nerves, demonstrating neurotropism, another form of EVMM. In addition, PM and EVMM of malignant tumor cells along the celiac trunk in patients with pancreatic carcinoma have been detected

[16,42]. The spread of pancreatic cancer along major vessels to sites remote from the primary pancreatic neoplasm was detected, that is, PM/EVMM (Fig. 4.12A–C). The presence of pancreatic carcinoma cells along the abluminal surfaces of the celiac trunk without intravasation was confirmed by endoscopic ultrasound fine-needle aspiration.

Finally, the presence of angiotropic tumor cells in human invasive prostate cancers, associated with PM of prostate cancer cells cultivated in vitro and in vivo, suggests that PM could be a mechanism of migration/metastatic spread in this malignancy as well [43]

Molecular studies

During glioma invasion, tumor cells progress along the abluminal vascular surfaces and such migration represents PM. It has been shown that bradykinin stimulates amoeboid migration of glioma cells [44]. In addition, CXCR4 signaling is also critical for perivascular invasion of glioma cells [45]. Among other molecules involved in glioma migration along vessels, Wnt7 signaling also stimulates tumor progression [46]. As mentioned earlier, PM has been shown to be a mechanism of metastatic colonization in multiple organs [26]. During PM, L1-CAM increases β1 integrin-ILK signaling for YAP nuclear localization [14]. In addition, CD44, which notably is a marker of pericytes and of CSCs in different types of cancers, promotes migration and invasion of docetaxel-resistant prostate cancer cells via induction of Hippo-YAP signaling [47]. In addition, the actin-related proteins 2/3 complex (Arp2/3 complex) mediate the nucleation of actin filaments at the leading edge of cells to drive cell movement, and has been previously implicated in the motility and invasion of both breast cancer cells and colorectal cancer cells [48]. Furthermore, ARPC3 expression is significantly higher in replacement HGP metastases when compared to desmoplastic HGP metastases [48].

Laminin and cancer cell migration

Laminins are a family of αβγ heterotrimeric glycoproteins usually present as the major component in the basement membrane. The structural and compositional "remodeling" of extracellular matrices and especially the basement membrane during several physiological and pathological events triggers new interactions with cell-surface receptors and cytoplasmic signaling pathways responsible of a variety of biological processes. Laminins are in particular implicated in cell adhesion, differentiation, migration, signaling, neurite outgrowth, and metastasis [49,50].

Laminins in melanoma

In melanoma, several studies have indicated a role for laminins in PM.

Angiotropism in melanoma has observed by electron-microscopy. At the ultrastructural level, endothelial cells appear intact. However, when melanoma cells interact with endothelial cells, an amorphous basement membrane zone is observed. This matrix appears to lack assembly into the usual organized supramolecular structure recognized as conventional basement membrane [51,52] (Fig. 4.10A and B). Such findings are not observable at the light microscopy level. Notably, during vascular co-option the presence of a preserved vascular basement membrane continuous with that of the nonmalignant tissue was determined by immunostaining [7], but the detection of a potential amorphous basement membrane may require ultrastructural analyses.

This amorphous basement membrane matrix contains mainly laminin, which could expose cryptic promigratory sites that trigger increased cellular migration [53], that is, PM. Indeed, laminins have been implicated in PM. For example, human melanoma cells spreading on the CAM (i.e., PM) overexpressed several laminin chains and laminin receptors [54]. Furthermore, peptide C16, a laminin $\gamma1$ chain synthetic peptide, had been shown to significantly promote cell adhesion, enhance pulmonary metastasis and migration of murine melanoma B-16 cells. C16 has also PM-promoting activity with human melanoma cells on the CAM [29]. Finally, melanoma cells produce several laminin isoforms and migrate on the $\alpha5$ laminin chain [55].

Laminins in non-melanoma tumors

Laminins are also implicated in several different cancer cell types [41]. Increase in laminin expression may facilitate PM along vessels since tumor cells associate with endothelial cells predominantly through laminin [49,50]. For example, laminin-332 is implicated in tumor migration, invasion, and metastasis. Interestingly, pulmonary laminin-332 may play a role in the pulmonary metastases of breast carcinoma [56].

In addition, the $\gamma2$ laminin chain, which contributes to cancer cell motility, is highly expressed by human cancers at the invasion front, and this expression correlates with poor cancer prognosis [57,58].

$\alpha4$-Laminins, such as laminins 411 and 421, are mesenchymal laminins expressed by blood and lymphatic vessels and some tumor cells [59]. During PM progression of glial tumors, laminin-421 is switched to laminin-411 ($\alpha4\beta1\gamma1$), which is dramatically increased in glial brain tumors [59]. In addition, there is a clinical correlation between the expression of laminin-411 with higher tumor grade and with expression of CSC markers, including notch pathway members, CD133, Nestin, and c-Myc [59].

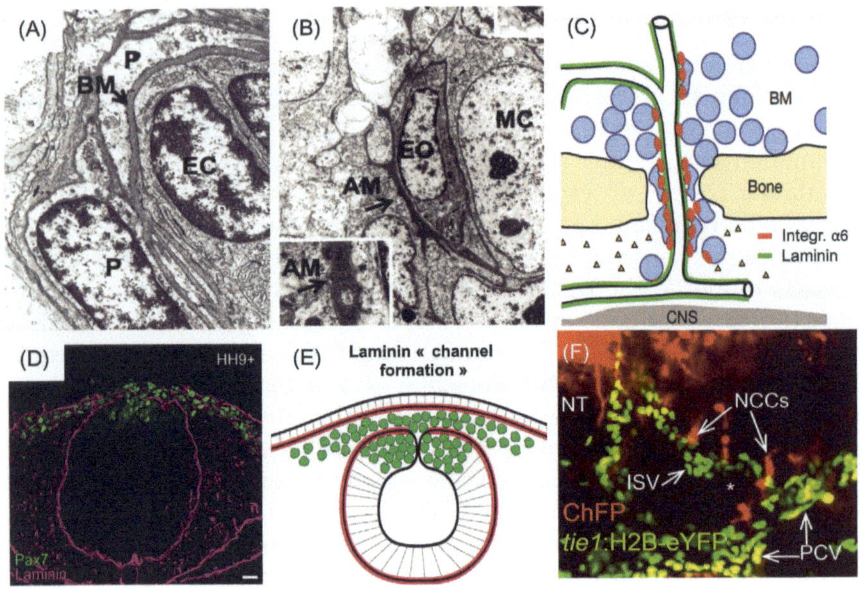

FIGURE 4.10 PM and laminin in cancer and embryogenesis.
(A and B) Pericytes versus melanoma cells. Electron microscopy showing an intradermal microvessel. The vessel consists of endothelial cells (EC) around a close lumen and pericytes (P). Note the complete basement membrane (BM) exhibiting a lamina densa and a lamina lucida. (B) Electron microscopy of a metastatic melanoma. Melanoma cells (MC) are closely associated to the abluminal surface of an endothelial cell (EC) from a microvessel (angiotropism). Note the amorphous matrix (AM), without the complete architecture of a stable basement membrane, around and between the two kinds of cells, which is detailed in the lower left inset. (C) Schematic of leukemic cells invasion of the central nervous system subarachnoid space via PM along the abluminal surface of laminin + bone marrow emissary vessels. (D and E) Cranial neural crest cells migrate through a laminin channel during EMT. The basement membrane is remodeled during cranial neural crest EMT. Immunostaining for neural crest cells (Pax7 green) and laminin (magenta) in cross sections of HH9 + embryos. Scale bar, 20 μm. (D) As Pax7 + neural crest cells undergo EMT, a laminin channel forms by HH9 +, creating a passage through which the cells are able to migrate. (E) Diagram of immunostaining data. *Green circles*: neural crest cells; *magenta lines*: laminin-rich basement membrane. (F) PM during neural crest development. Neural crest cells (NCCs) track along the developing intersomitic vessel (ISV) to colonize the sympathetic ganglia. *NT*, neural tube; *PCV*, postcardinal vein. *Sources: A–B from Ref. [51]; C from Ref. [15]; D–E from Ref. [60]; F from Ref. [78].*

In leukemia, the interaction of α6 integrin from leukemia cells with abluminal vascular laminin is involved in the invasion in the central nervous system via PM along bridging vessels [15] (Fig. 4.10C).

Finally, α5 chain-containing laminins are involved in cell–cell adhesion and migration in several malignancies, in particular via Lu/BCAM receptor [60]. Notably, modulation of laminin plays an essential role in regulating cranial NC migration, in particular laminin α5 [60].

Modulation of laminins in embryogenesis and in cancer progression

Embryonic cells and CSCs proactively remodel their microenvironment to maintain their progression programs [61,62]. Notably, laminin is the main component of the basement membrane matrix present in many embryonic tissues, and during embryonic migration, specific laminins are expressed by the basal surfaces of the epithelia/endothelia lining these pathways [50]. By remodeling the extracellular matrix and adhesion proteins, embryonic cells and CSCs migrate to distant sites through tissues. Such modifications affect intracellular signaling pathways by exposing or cleaving latent proteins and/or factors trapped in the extracellular matrix, and in turn affect the cytoskeleton. It may therefore be more relevant to speak of "remodeling" rather than "degradation" of the extracellular matrix and basement membrane matrix during embryogenesis and cancer progression.

PM and EVMM: Reversion to an embryogenesis-derived program

1. Analogies between embryogenesis and cancer development
2. Neural crest cell migration: a model for PM/EVMM
3. Angiogenesis.
4. Cell competition.
5. PM and EVMM: routes, direction and timing according to an embryogenesis-derived program

Analogies between embryogenesis and cancer development

The close relationship between the processes involved in both tumorigenesis and development of the embryo date back to 1859 when Rudolf Virchow indicated that "neoplasms arise in accordance with the same laws that regulate embryonic development" [63]. It has been shown that embryogenesis involves major morphological changes and cellular migrations that are imitated/repeated in cancer [64]. In addition, reactivation of embryonic signals in adult cells, as a consequence of mutations and epigenetic remodeling, is a characteristic feature of cancer [65].

- Epithelial-to-mesenchymal transition

The EMT describes a cellular process during which epithelial cells transition to a mesenchymal cell state. EMT is critical during embryonic development and organogenesis [66]. During EMT, epithelial cells (or intraepithelial cells, such as melanoblasts) lose their epithelial characteristics, including both apical-basal polarity and cell–cell adhesion, and acquire migratory and invasive mesenchymal cell capabilities to escape from their tissue of origin. Similarly, EMT plays a role during cancer progression. In addition, its reverse process, a mesenchymal to epithelial transition (MET), seems to support metastatic outgrowth inside distant organs [66]. EMT in metastasis, like in embryogenesis, may result in a spectrum of phenotypes between the "absolute" epithelial and mesenchymal cell states [66].

- Cancer stem cells

Many tumors contain a subpopulation of tumor cells with stem/embryonic cell features, termed as CSCs [67]. This population is highly plastic with bidirectional interconversions between stem and nonstem states via an EMT to MET switch. CSCs have self-renewal, migration, metastasizing, and drug resistance properties [68]. During PM, the interaction of malignant cells with the abluminal vascular surface induces the stem-embryonic-like properties of tumor cells [13]. Notably, tumor cells in PM replace pericytes which are mesenchymal stem cells [69].

- Embryonic gene expression in PM

Many human tumors recapitulate the early gene expression of embryogenesis [64]. It is also suggested that the aggressiveness capacities of cancer depend on the embryonic layer of origin [64]. For example, key developmental signaling pathways, including the Wnt, Hedgehog, and Notch pathways, are frequently dysregulated in cancer and participate in all stages of tumor progression, from initiation and maintenance to metastatic spread and growth at distant sites [65]. As described above, differentially expressed genes during melanoma PM in vitro have demonstrated properties, among others, linked to EMT, such as embryonic and/or CSC properties [34]. In addition, the 15 most significantly enriched functional groups analyzed by Ingenuity-Pathway-Analysis included development, cell movement, cancer, and embryonic development [34]. Finally, from the microarray analysis of melanoma cells exhibiting PM, several differentially expressed genes were highlighted that are present in the human embryonic stem cell pluripotency pathway, including TGFβ/NODAL, PI3K, SMAD, and EDG, which are directly implicated in self-renewal [34]. These data suggest that the interaction between angiotropic melanoma and endothelial cells upregulates specific embryonic/mesenchymal stemness programs during PM.

- Absence of intravascular cell circulation in embryogenesis

In the human embryo, "the dogma has been that maternal blood flows through the developing placenta from days 17–20 onward, thus establishing the hemotrophic pathway" [70]. However, it is known today that histiotrophic nutrition is an essential source of nutrients during organogenesis, and that the maternal arterial circulation to the human placenta is not fully established before the end of the first trimester [71]. Therefore despite all these recognized analogies between tumor progression and organogenesis, it has to be emphasized that the intravascular cell spread described in tumor metastasis does not exist during the course of organogenesis. In addition, is it "biologically coherent" for a plastic and ultra-sensing tumor cell to leave a vascular niche with oxygen and nutrients, allowing migration and proliferation, and thus not experiencing hemodynamic shear forces which will almost certainly destroy it? Indeed, intravascular metastasis is an "inefficient process" with most CSCs dying within the bloodstream [72]. Could the *uniqueness* of the "inefficient" intravascular cancer cell dissemination pathway be a "dogma," therefore banning all forms of alternative...? [73].

- Seed-and-soil paradigm

The seed-and-soil paradigm of Paget stated in 1889 that "tumor cells (the seeds) have a specific affinity for specific organs (the soil), and metastasis does not occur by chance" [74]. Since it has been verified that most malignant tumors show an organ-specific pattern of metastasis [75]. It should be noted that during organogenesis, embryonic stem cells are unquestionably the "seeds" migrating through the embryo to reach their ultimate phenotype-specific sites or "soil."

Neural crest cell migration: a model for PM/EVMM

- Neural crest cell migration

There are strong analogies between PM/EVMM and the mechanisms of migration during embryogenesis, particularly during the migration of neural crest cells in the developing embryo. The neural crest is a highly migratory embryonic cell population that develops into numerous cell lineages, including melanocytes, smooth muscle, neurons, and craniofacial mesenchyme. During EMT, polarized premigratory neural crest cells in contact with a basement membrane matrix lose apical-basal polarity to delaminate from the dorsal neural tube. Notably, cranial neural crest migrates through a laminin channel during EMT (Fig. 4.10D and E). Subsequently, neural crest cells migrate along specific routes to a number of sites where the cells arrest and fully differentiate into a wide variety of derivatives [76,77]. Notably, neural crest cells track alongside developing intersomitic vessels at multiple locations along their trajectories to colonize the sympathetic ganglia, but never cross the continuous streams of endothelial cells [78] (Fig. 4.10F). This mechanism of neural crest cell migration is very similar to that of tumor cells during PM/EVMM.

- Embryonic program of melanoma cells

Human metastatic melanoma cells, when transplanted into the chick embryonic neural crest microenvironment, respond to cues from the surrounding host tissues and emigrate along stereotypical migratory routes traveled by neural crest cells [79]. Furthermore, "melanoma revives an embryonic migration program to promote plasticity and invasion" and aberrant regulation of neural crest developmental genes may promote melanoma plasticity and invasiveness [80].

- Embryonic program of glioma

The highly invasive nature of human gliomas may recapitulate the migratory behavior of glial progenitors, that is, astrocytes and oligodendrocytes. Oligodendrocyte precursors migrate along the vasculature in the developing nervous system, that is, PM/EVMM [81] (Fig. 4.11A and B). Notably, the migratory properties of glioma cells strongly resemble that of glial progenitors [82]. Furthermore, in an in vivo model of brain tumors, stem-like glioma cells migrating along vessels expressed neural stem cell markers and showed a migratory behavior similar to that of normal human neural stem cells [83].

- Neural crest cells migrate as a continuous mass of tissue along the neural tube and subsequently split into spatially distinct migratory streams to invade/spread to precise parts of the embryo [83,84]. Neural crest cell migration can be described as follows: "Once neural crest differentiation occurs, cells occupying the dorsal portion of the neural tube lose their apico-basal polarity and disrupt cell junctions, acquire EMT and motile properties, migrate away from the neural tube and embark upon an extensive continuous migration through the embryo to reach their ultimate phenotype-specific sites without entering in vascular lumen." Melanoma is known to have very similar invasive ability as that of neural crest cells [85], and we propose that it is possible to describe melanoma cell metastasis as follow:

"Once a solid tumor becomes invasive, cells occupying the advancing front of the tumor lose their apico-basal polarity and disrupt cell junctions, acquire EMT and motile properties, migrate away from the primary tumor, and embark upon an extensive continuous migration through the body to reach their metastatic secondary sites without entering the vascular lumen," that is, EVMM.

Given the reactivation of embryonic signals in cancer [86], it may be interesting to consider the similarities between EVMM in cancers with

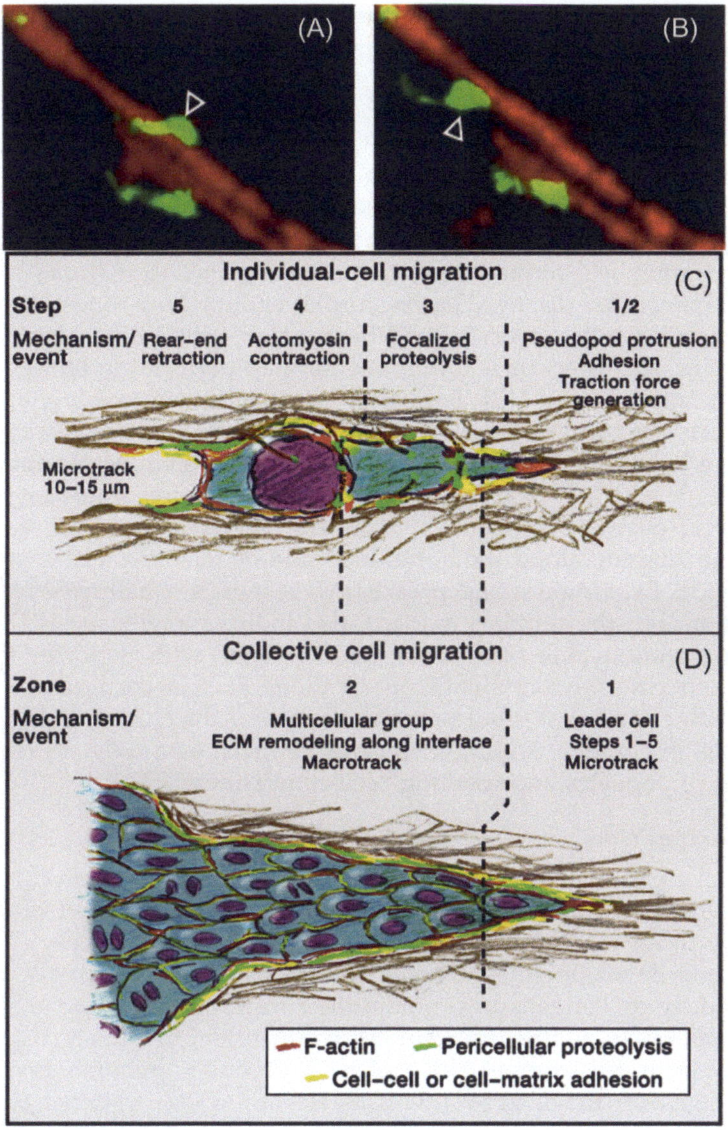

FIGURE 4.11 Cell migration in embryogenesis and tumor progression.
(A and B) Time-lapse imaging of oligodendrocyte precursor cells (OPCs) crawling along a penetrating vessel in the E18 cortex. (C) Single-cell migration involves five molecular steps that change the cell shape, its position, and the tissue structure through which it migrates. (D) Collectively migrating cells form two major zones: zone 1, in which a "leader cell" generates a proteolytic microtrack at the front of the migrating group, and zone 2, in which the subsequent cells then widen this microtrack to form a larger macrotrack. Note the modulation of the extracellular matrix (ECM). *Source: (A and B) Modified from Ref. [81]; C–D from Ref. [94].*

different embryonic origins and the respective pathways of their progenitors during embryonic development.

Angiogenesis

Early in embryonic development, vessel formation occurs by a process referred to as vasculogenesis in which endothelial cells differentiate and proliferate in situ within a previously avascular tissue. In cancer biology, the simulation of this event by cancer cells that form channels within tumors was termed *vasculogenic mimicry* in 1999 [87], *and described in depth in another chapter*. The primordial endothelium, once assembled into vascular tubes, recruits undifferentiated cells with mesenchymal properties and direct their differentiation into pericytes in both capillaries and small vessels [13]. Formation of a basement membrane matrix completes the process. Through these mechanisms, a circulatory system is formed and remodeled into a complex vessel system before any blood circulation which appears only at the end of the first trimester [88,89]. Similarly, during angiogenesis in adults, pericytes are recruited and begin to migrate along the abluminal side of microvessels to stabilize them [13]. Expansion along preexisting vessels is a hallmark of vessel co-option, but the capillary pattern also induces angiogenesis [7]. It is therefore conceivable that tumor cells compete with pericytes for PM along neovessels. Experimental observations, such as coculture of tumor cells with endothelial tubules, aortic rings, and the chick CAM [8], support this hypothesis. During PM, invasive melanoma cells are recruited instead of pericytes, representing "cell competition."

Cell competition

Cell competition is a fitness-sensing embryonic mechanism conserved from Drosophila to vertebrates resulting in the replacement of less fit cells by those with higher fitness. Cell competition is not only essential for proper development and tissue homeostasis in adult, but also can be hijacked by cancer cells to promote their transformation and their progression [90]. In nonneutral stem cell competition, two unequal populations of stem cells compete for a niche, with one population gaining an advantage over the other [91]. This is relevant to PM, where angiotropic CSCs are competing with marrow stromal pericytes for the vascular niche. There is probably no "competition" for intravasation since the blood circulation is such an unfavorable environment for tumor cells [17,72].

PM and EVMM: routes, direction and timing according to an embryogenesis-derived program

During intravascular dissemination, most CSCs are rapidly eliminated before being able to form a metastasis, which explains why the intravascular metastatic process was termed "inefficient" [17,72]. In metastases, contact of tumor cells with vessel lumina (angiotropism/co-option) is considered as an event following extravasation after the "express" dissemination (fraction of a second) of CSCs. Therefore almost all metastatic experimental models use cancer cells directly injected into the circulation to analyze the resulting metastases, consequently bypassing the real metastasis timing and pathways from the primary to the metastatic sites. EVMM may represent an embryogenesis-derived metastatic pathway in which tumor cells adhere to a programmed migration from a primary site to more distant metastatic niches (Fig. 4.1). EVMM could be an additional and/or a more specifically cancer-related mechanism, while intravascular dissemination, which is not derived from an embryonic program, may be a passive or "accidental" event during the cancer invasion of the surrounding normal tissue. Further studies are needed to investigate the relative frequency of these two different mechanisms. The process of embryogenesis requires precise spatial and temporal activation of developmental signaling pathways [65]. During EVMM, cancer progression and migration may aberrantly recapitulate embryonic migration in a mature and therefore unintelligible environment. This view is supported by the observation that tumor cells form a "normal" mosaic organism in an embryonic and therefore intelligible environment [73,86] In the following paragraphs, potential routes, direction and timing of EVMM are described in accordance with knowledge of embryogenesis.

Space, direction, and timing

- Space: the routes of migration

As described in neural crest migration in the embryo, tumor cells could migrate along stereotypic pathways to numerous distant sites [13,77]. In addition to the replacement of pericytes along microvessels, EVMM can also involve other tracks or pathways, as described in embryogenesis [77]. Similarly [67], cancer cells may compete with other perivascular mesenchymal stem cells such as adventitial cells [90,91]. Notably, by extension the term PM represents migration along vessels of all types.

Celiac axis malignant perivascular cuffing and EVMM have been detected in patients with metastatic pancreatic cancer [16,42]. Here cancer cells may compete with either mesenchymal stem cells or adventitial cells located in large vessels [69].

A variety of other anatomic structures are used as a scaffolding for cancer cell migration, sometimes for considerable distances. Such scaffolding includes nerves (neurotropism or perineural invasion) [93], the peritoneum, pleura, myofibers, adipocytes, bone cavities, brain ventricles, choroid plexus, and the glia limitans in the brain [94]. Furthermore, remodeling of the extracellular matrix by cancer-associated fibroblasts generates tracks for cancer cells [95].

Thus via a combination of PM and other extravascular routes, EVMM could involve a progressive migration of cancer cells to both nearby and more distant sites.

- Direction

During organogenesis and after EMT, a front-rear polarity axis of stem cells will trigger a regulated directional migration under physical and chemical guidance cues to reach their correct destinations and differentiate into appropriate derivatives [77]. Directional migration arises as a consequence of either single or combinatorial factors (haptotaxis, durotaxis, and anisotropic organization of the extracellular matrix, etc.) [97]. Cancer cells also use contact guidance and chemotactic cues for reaching their final metastatic site. Cancer cell migration is initiated and maintained by signaling pathways that control cytoskeletal dynamics in tumor cells and the turnover of cell–matrix and cell–cell junctions, followed by cell migration into the adjacent tissue [88]. Cells can migrate individually or collectively as multicellular groups (Fig. 4.11C and D). Collective migration is potentially the predominant form of metastatic migration, and neoplastic collective movements are likely to recapitulate morphogenic movements [97], however, without possible proper regulation.

The interaction of the primary tumor with the environment of distant organs may result in a premetastatic niche, during which extracellular matrix remodeling would be an essential process [98]. Interestingly, laminin-332 from lung epithelium may induce breast cancer cell migration to the lungs [51,56]. In addition, CCL2 and IL6, which are overexpressed in a 3D model of PM [34], are implicated in promoting organ-specific metastasis [99]. Despite and beside organotropism, random and dysregulated directional cancer migration takes place during cancer progression. This absence of specificity may be due to the "unintelligible" mature microenvironment in which cancerous cells repeat indefinitely the mechanisms of migration and growth until the exhaustion of the invaded organ.

- Time

Cancer dormancy is a term used to explain the extensive period of time in which patients remain asymptomatic prior to relapse [100]. Dormancy, which refers to the presence in metastatic niches of dormant CSCs/circulating tumor cells after extravasation, is a loosely defined phenomenon that remains poorly understood [100,101].

Interestingly, long-term dormancy periods can correspond to angiotropic cancer cells migrating continuously towards the metastatic sites [102], i.e. PM/EVMM. Therefore, EVMM which is a slow and continuous process, can provide an alternative explanation for the delayed formation of metastases in many cancer cases. How long would it take for the tumor cells to reach secondary sites without intravascular dissemination? Giving cell velocity in distance per year can help define the time needed by tumor cells for reaching secondary sites via PM/EVMM, that is, without intravascular dissemination. Migration modes and dynamics can be observed by intravital imaging of invasive human tumor cells [103]. Cells with ameboid-like morphology tend to migrate either as single cells or as streams, while cells with predominantly mesenchymal phenotypes are switching between single-cell and collective migration modes. For example, the in vivo velocity (in cm per year) of human melanoma varies from 22 to 525 cm/year, and human breast cancer cells migrate from 13 to 220 cm/year [103]. These speeds are theoretically compatible with the long-time intervals between the recognition of the primary cancer and the formation of metastases and could be applicable to PM/EVMM. Notably, neural crest cell migration rates are comparable to migrating tumor cells. Indeed, the average velocity of neural crest cells in the zebrafish embryo is 40 cm/year [104], and 46 cm/year in the chick embryo [105]. These long-time periods for PM/EVMM may explain in part the dormancy period observed with many tumors.

In summary, given (1) the well-recognized analogies between embryonic and cancer cell migration, (2) the compatibility between embryonic and cancer cell migratory pathways and velocities, (3) the absence of circulation during embryonic organogenesis, and therefore migration only via PM/extravascular migration, all in concert support the concept that these extravascular embryonic migratory mechanisms abnormally recur during cancer metastasis. Via PM/EVMM, cancer cells will follow indefinitely, in a dysregulated way, "embryonic seed and soil" mechanisms until the exhaustion of the invaded organ.

Perspectives and therapeutic implications

Detection of PM and EVMM

"What man sees depends both upon what he looks at and also upon what his previous visual-conception experience has taught him to see" [1].

Angiotropism/vessel co-option is histopathologically observed in tissue samples. Since PM/EVMM are not (yet) part of the "normal science," methods for analyzing these phenomena have not been explored until recently. For example, further study on PM and EVMM would benefit from the detection of a distance traveled by many tumor types. The detection of PM/EVMM at distances from a primary pancreatic cancer has been recently realized safely and with accuracy via endoscopic ultrasound fine-needle aspiration [16,42] (Fig. 4.12A–C). Such detection has been shown to affect treatment planning, by increasing

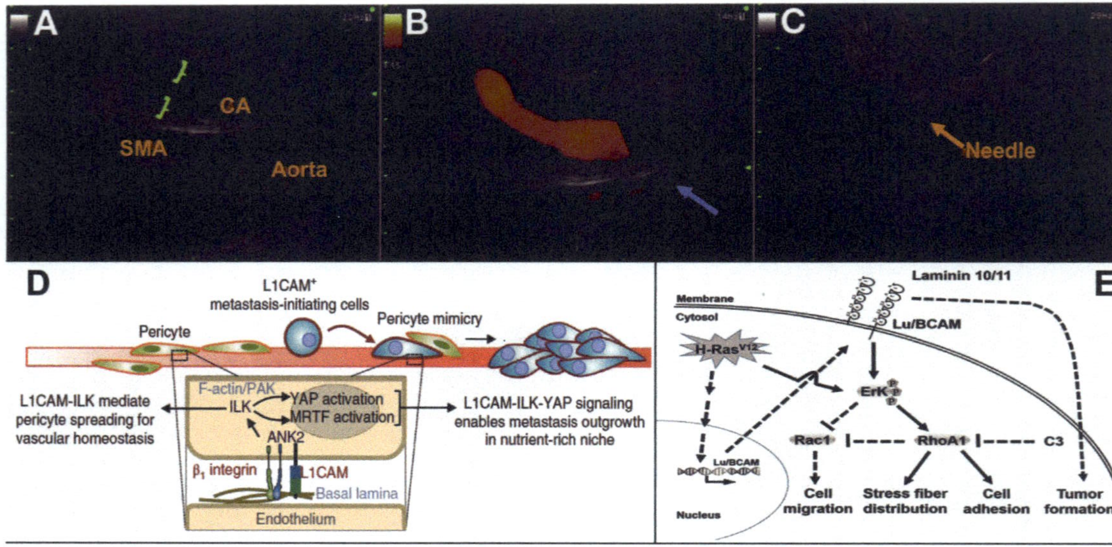

FIGURE 4.12 Detection and therapeutic perspectives of PM/EVMM. (A–C) Endoscopic ultrasound (EUS) needle aspiration detection of extravascular migratory metastasis from a remotely located pancreatic cancer. (A) In a patient with a 25–20 mm pancreatic body mass that appeared confined to the pancreas on computed tomography, EUS showed an irregular hypoechoic process (*green brackets*) of varying thickness surrounding the celiac artery. (B) Power Doppler shows subtle narrowing of the celiac artery as it courses through the hypoechoic vascular cuffing. (C) Under EUS guidance, the site of greatest thickness was targeted with fine-needle aspiration (*arrow*), showing the presence of adenocarcinoma. (D) Model of PM mechanism: L1CAM-mediated pericyte-like spreading (PM) and subsequent β1 integrin, ILK and YAP activation adapted by metastatic cells for colonization. (E) A hypothetical model of Lu/BCAM-laminin-mediated F-actin rearrangement, cell adhesion, migration and tumor formation during bladder tumorigenesis. Under Lu/BCAM overexpression conditions in the presence of laminin, the Erk/MAPK signaling pathway is activated, which is responsible for activation of RhoA and suppression of Rac-1. Subsequently, F-actin rearrangement and cell adhesion are induced, and cell migration was suppressed. *Sources: A–C from Ref. [42]; D from Ref. [14]; E from Ref. [110].*

the accuracy of tumor staging, determination of tumor resectability, and therapeutic options. It would be very useful to investigate the use of endoscopic ultrasound fine-needle aspiration in other cancers for assessing EVMM and improving the accuracy of cancer staging. In addition, such a technique could be invaluable for studying migrating tumor cells.

Therapeutic perspectives

Cancer cells committed in PM are progressing along a continuum of vascular niches and have stem-embryonic-like features with genetic, epigenetic, and cellular adaptations conferring them resistance to classical therapeutic approaches, in particular to antiangiogenic therapy [101,106].

- Inhibiting vessel co-option/angiotropism

Several strategies have been suggested for inhibiting vessel co-option/angiotropism reviewed in [7] and in Chapter 3, Vascular co-option. In particular, strategies against tumor migration and vascular adhesion are relevant to PM.

- Targeting molecules linked to embryonic development and cancer migration

Studying embryonic factors linked to migration and detected during cancer metastasis may be particularly relevant to PM/EVMM.

DAN

DAN, a factor which inhibits uncontrolled neural crest and metastatic melanoma invasion, has been suggested as a candidate inhibitor of metastatic melanoma [85].

YAP signaling

YAP enhances human neural crest cell migration [107], and therapeutics targeting YAP activation are increasingly considered for development. As mentioned above, PM has been shown to be a mechanism of metastatic colonization in multiple organs [14]. During PM, L1CAM increases β1 integrin-ILK signaling for YAP nuclear localization (Fig. 4.12D). These data could lead to future therapeutic targeting of this signaling pathway. In addition, CD44, which notably is a marker of pericytes and of CSCs in different types of cancers [108], promotes migration and invasion of docetaxel-resistant prostate cancer cells via induction of Hippo-Yap signaling which may be a target in this neoplasm [109].

Laminins

The role of laminins in embryogenesis and tumor migration could be similar, and this could have important therapeutic implications. Increasing studies document the great potential of laminins and their receptors as (1) biomarkers for prognosis or prediction of treatment response and (2) valuable targets for therapeutic interventions against CSC migration and metastasis [50]. Some examples:

(1) LM-411 expression in the glioma microenvironment correlates with Notch and with other CSC markers and can be targeted by a novel, clinically translatable nanobioconjugate to inhibit glioma growth [59]. (2) Targeting the laminin receptor α6 integrin on leukemia cells prevents the interaction between vascular laminin and tumor α6 integrin during invasion of the central nervous system via PM [15] (Fig. 4.10C). Either a PI3Kδ inhibitor or specific α6 integrin-neutralizing antibodies are considered for therapeutic tests [15]. (3) Laminin-332 plays a role in the pulmonary metastases of breast carcinoma and may provide a target for antimetastatic therapy [56]. (4) Lu/BCAM functions as the receptor for specific ligand laminin-α5 chain and is involved in cell–cell adhesion and migration in several malignancies [110]. This laminin receptor may have potential as a novel therapeutic target [110,111] (Fig. 4.12E).

Concluding remarks

PM and vascular co-option: complementary embryonic phenomena?

PM is a step-by-step process of cancer cell migration along vascular channels.

This perivascular location of tumor cells intrinsically links PM to vascular co-option, possibly via a reversible "phenotype-switch" involving EMT. In the embryo, when subpopulations of cells are programmed to travel long distances to peripheral tissue sites, cell division and migration must be tightly integrated [112]. During tumor progression, EMT leads to a phenotypic change from a proliferative to an invasive, migratory state. This principally reversible "phenotype-switch" model of metastasis may be driven by BRN2 in melanoma [113].

It has been proposed that a switch from vessel co-option to angiogenesis can occur during tumor progression [7]. Here, we propose that adhesion to and spreading on the vascular basement membrane matrix modulate this basement membrane, triggering reversible tumor cell switches via dysregulated embryonic mechanisms: (1) from a migratory phenotype, that is, "migration of tumor cells to reach their (distant)

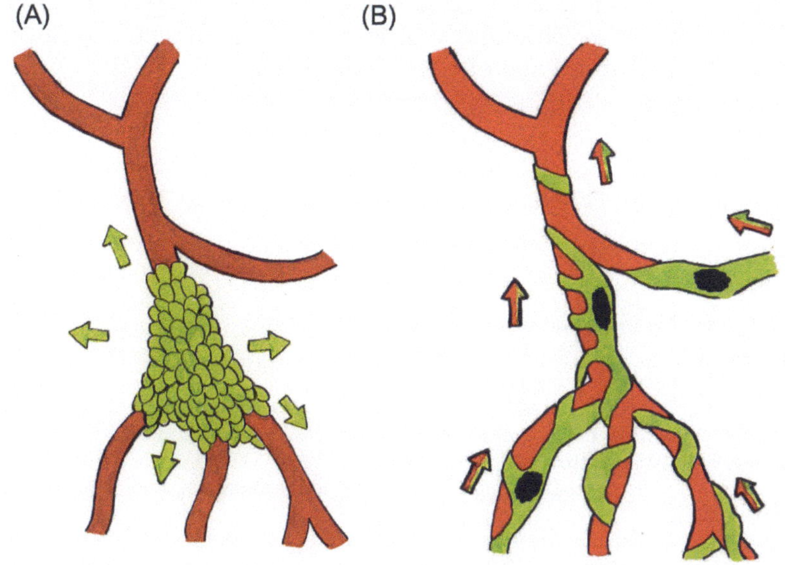

FIGURE 4.13 Diagram of vascular co-option and PM.
(A) Vascular co-option. Note the tumor growth via local tumor invasion and proliferation along co-opted vessels. (B) PM. Note the spread of tumor cells via continuous migration along the abluminal vascular surfaces (to nearby and potentially more distant sites).

secondary sites" during PM/EVMM and (2) to an invasive/proliferative phenotype during co-option for "colonizing their primary and secondary sites" during vascular co-option (Fig. 4.13). The replacement pattern in liver metastasis may represent both vessel co-option for blood supply and tumor growth [7], and PM for migration to secondary sites (Figs. 4.1D and E and 4.6). Indeed, the spreading tumor cells may arrive in the liver via PM (EVMM), invade the liver by a co-option/PM switch, and may exit this organ in route to other secondary sites again by way of PM/EVMM, realizing a continuous migration throughout the body (Fig. 4.1).

Finally, according to governmental agencies and industry, for over six decades cancer treatments have had only limited success and for only short periods of time, and many failures have been experienced [114]. The long interval to recurrence after resection and after other therapies of various types have proved very troublesome and, particularly following many therapies, the development of metastatic tumors that are resistant to therapy. The phenomenon of dormancy remains poorly understood at best, and, accordingly, therapeutic interventions have not been formulated. A comprehensive understanding of the mechanism(s) of tumor spread is needed. Vascular co-option and PM/EVMM are new paradigms that may profoundly influence and stimulate the development of more powerful and effective therapeutics.

Acknowledgments

We sincerely thank Maud Haon who designed the diagrams for Figs. 4.1 and 4.13.

References

[1] Kuhn TS. The structure of scientific revolutions. Chicago, IL: University of Chicago Press; 1962.
[2] NCI Dictionary of Cancer Terms.
[3] Chitty JL, Filipe EC, Lucas MC, Herrmann D, Cox TR, Timpson P. Recent advances in understanding the complexities of metastasis. Version 2. F1000Res 2018;7 pii: F1000 Faculty Rev-1169.
[4] Folkman J. Tumor angiogenesis: therapeutic implications. N Engl J Med 1971;285 (21):1182–6.
[5] Weidner N, Folkman J. Tumoral vascularity as a prognostic factor in cancer. Important Adv Oncol 1996;167–90.
[6] Barnhill RL, Busam KJ, Berwick M, Blessing K, Cochran AJ, Elder DE, et al. Tumour vascularity is not a prognostic factor for cutaneous melanoma. Lancet 1994;344 (8931):1237–8.
[7] Kuczynski EA, Vermeulen PB, Pezzella F, Kerbel RS, Reynolds AR. Vessel co-option in cancer. Nat Rev Clin Oncol 2019;16(8):469–93.
[8] Lugassy C, Zadran S, Bentolila LA, Wadehra M, Prakash R, Carmichael ST, et al. Angiotropism, pericytic mimicry and extravascular migratory metastasis in melanoma: an alternative to intravascular cancer dissemination. Cancer Microenviron 2014;7(3):139–52.
[9] Lugassy C, Escande JP. The haematogenous theory of metastasis: récamier did not propose it. Virchows Arch 1997;431(5):431.
[10] Barnhill RL, Lugassy C. Angiotropic malignant melanoma and extravascular migratory metastasis: description of 36 cases with emphasis on a new mechanism of tumour spread. Pathology 2004;36(5):485–90.
[11] Lugassy C, Eyden BP, Christensen L, Escande JP. Matrix interactions between tumor cells and endothelium in human malignant melanoma. J Invest Dermatol 1996;106:894.
[12] Lugassy C, Barnhill RL, Christensen L. Melanoma and extravascular migratory metastasis. J Cutan Pathol 2000;27(9):481.
[13] Lugassy C, Péault B, Wadehra M, Kleinman HK, Barnhill RL. Could pericytic mimicry represent another type of melanoma cell plasticity with embryonic properties? Pigment Cell Melanoma Res 2013;26(5):746–54.
[14] Er EE, Valiente M, Ganesh K, Zou Y, Agrawal S, Hu J, et al. Pericyte-like spreading by disseminated cancer cells activates YAP and MRTF for metastatic colonization. Nat Cell Biol 2018;20(8):966–78.
[15] Yao H, Price TT, Cantelli G, Ngo B, Warner MJ, Olivere L, et al. Leukaemia hijacks a neural mechanism to invade the central nervous system. Nature 2018;560 (7716):55–60.
[16] Rustagi T, Gleeson FC, Chari ST, Lehrke HD, Takahashi N, Malikowski TM, et al. Safety, diagnostic accuracy, and effects of endoscopic ultrasound fine-needle

aspiration on detection of extravascular migratory metastases. Clin Gastroenterol Hepatol 2019; pii: S1542-3565(19)30353-2.

[17] Talmadge JE, Fidler IJ. AACR centennial series: the biology of cancer metastasis: historical perspective. Cancer Res 2010;70(14):5649–69.

[18] Ferland-McCollough D, Slater S, Richard J, Reni C, Mangialardi G. Pericytes, an overlooked player in vascular pathobiology. Pharmacol Ther 2017;171:30–42.

[19] Barnhill RL, Lugassy C. Angiotropic malignant melanoma and extravascular migratory metastasis: description of 36 cases with emphasis on a new mechanism of tumour spread. Pathology 2004;36(5):485–90.

[20] Barnhill RL, Ye M, Batistella A, Stern MH, Roman-Roman S, Dendale R, et al. The biological and prognostic significance of angiotropism in uveal melanoma. Lab Invest 2017. Available from: https://doi.org/10.1038/labinvest.2017.16.

[21] Barnhill RL, Lemaitre S, Lévy-Gabrielle C, Rodrigues M, Desjardins L, Dendale R, et al. Satellite in transit metastases in rapidly fatal conjunctival melanoma: implications for angiotropism and extravascular migratory metastasis (description of a murine model for conjunctival melanoma). Pathology 2016;48(2):166–76.

[22] Barnhill R, Vermeulen P, Daelemans S, van Dam PJ, Roman-Roman S, Servois V, et al. Replacement and desmoplastic histopathological growth patterns: a pilot study of prediction of outcome in patients with uveal melanoma liver metastases. J Pathol Clin Res 2018;4(4):227–40.

[23] Bald T, Quast T, Landsberg J, Rogava M, Glodde N, Lopez-Ramos D, et al. Ultraviolet-radiation-induced inflammation promotes angiotropism and metastasis in melanoma. Nature 2014;507(7490):109–13.

[24] Van Es SL, Colman M, Thompson JF, McCarthy SW, Scolyer RA. Angiotropism is an independent predictor of local recurrence and in-transit metastasis in primary cutaneous melanoma. Am J Surg Pathol 2008;32(9):1396–403.

[25] Moy AP, Duncan LM, Muzikansky A, Kraft S. Angiotropism in primary cutaneous melanoma is associated with disease progression and distant metastases: a retrospective study of 179 cases. J Cutan Pathol 2019;46(7):498–507.

[26] Lugassy C, Kleinman HK, Fernandez PM, Patierno SR, Webber MM, Ghanem G, et al. Human melanoma cell migration along capillary-like structures in vitro: a new dynamic model for studying extravascular migratory metastasis. J Invest Dermatol 2002;119(3):703–4.

[27] Fornabaio G, Barnhill RL, Lugassy C, Bentolila LA, Cassoux N, Roman-Roman S, et al. Angiotropism and extravascular migratory metastasis in cutaneous and uveal melanoma progression in a zebrafish model. Sci Rep 2018;8(1):10448.

[28] Cheng L, Huang Z, Zhou W, Wu Q, Donnola S, Liu JK, et al. Glioblastoma stem cells generate vascular pericytes to support vessel function and tumor growth. Cell 2013;153:139–52.

[29] Lugassy C, Kleinman HK, Vernon SE, Welch DR, Barnhill RL. C16 laminin peptide increases angiotropic extravascular migration of human melanoma cells in a shell-less chick chorioallantoic membrane assay. Br J Dermatol 2007;157:780–2.

[30] Bentolila NY, Barnhill RL, Lugassy C, Bentolila LA. Intravital imaging of human melanoma cells in the mouse ear skin by two-photon excitation microscopy. Methods Mol Biol 2018;1755:223–32.

[31] Bentolila LA, Prakash R, Mihic-Probst D, Wadehra M, Kleinman HK, Carmichael TS, et al. Imaging of angiotropism/vascular co-option in a murine model of brain melanoma: implications for melanoma progression along extravascular pathways. Sci Rep 2016;6:23834.

[32] Barnhill R, Dy K, Lugassy C. Angiotropism in cutaneous melanoma: a prognostic factor strongly predicting risk for metastasis. J Invest Dermatol 2002;119(3):705–6.

[33] Lugassy C, Lazar V, Dessen P, van den Oord JJ, Winnepenninckx V, Spatz A, et al. Gene expression profiling of human angiotropic primary melanoma: selection of 15 differentially expressed genes potentially involved in extravascular migratory metastasis. Eur J Cancer 2011;47(8):1267–75.

[34] Lugassy C, Wadehra M, Li X, Corselli M, Akhavan D, Binder SW, et al. Pilot study on "pericytic mimicry" and potential embryonic/stem cell properties of angiotropic melanoma cells interacting with the abluminal vascular surface. Cancer Microenviron 2013;6(1):19–29.

[35] Valiente M, Obenauf AC, Jin X, Chen Q, Zhang XH, Lee DJ, et al. Serpins promote cancer cell survival and vascular co-option in brain metastasis. Cell 2014;156 (5):1002–16.

[36] Sweeney M, Foldes G. It takes two: endothelial-perivascular cell cross-talk in vascular development and disease. Front Cardiovasc Med 2018;5:154.

[37] Wesseling P, Ruiter DJ, Burger PC. Angiogenesis in brain tumors: pathobiological and clinical aspects. J. Neurooncol 1997;32:253–65.

[38] Gritsenko P, Leenders W, Friedl P. Recapitulating in vivo-like plasticity of glioma cell invasion along blood vessels and in astrocyte-rich stroma. Histochem Cell Biol 2017;148(4):395–406.

[39] Lugassy C, Haroun RI, Brem H, Tyler BM, Jones RV, Fernandez PM, et al. Pericytic-like angiotropism of glioma and melanoma cells. Am J Dermatopathol 2002;24(6):473–8.

[40] Beauchesne P. Extra-neural metastases of malignant gliomas: myth or reality? Cancers (Basel). 2011;3(1):461–77.

[41] Tiwary S, Morales JE, Kwiatkowski SC, Lang FF, Rao G, McCarty JH. Metastatic brain tumors disrupt the blood-brain barrier and alter lipid metabolism by inhibiting expression of the endothelial cell fatty acid transporter Mfsd2a. Sci Rep 2018;8 (1):8267.

[42] Levy MJ, Gleeson FC, Zhang L. Endoscopic ultrasound fine-needle aspiration detection of extravascular migratory metastasis from a remotely located pancreatic cancer. Clin Gastroenterol Hepatol 2009;7(2):246-–8.

[43] Lugassy C, Vernon SE, Warner JW, Le CQ, Manyak M, Patierno SR, et al. Angiotropism of human prostate cancer cells: implications for extravascular migratory metastasis. BJU Int 2005;95(7):1099–103.

[44] Seifert S, Sontheimer H. Bradykinin enhances invasion of malignant glioma into the brain parenchyma by inducing cells to undergo amoeboid migration. J Physiol. 2014, Nov 15;592(22):5109–27.

[45] Yadav VN, Zamler D, Baker GJ, Kadiyala P, Erdreich-Epstein A, DeCarvalho AC, et al. CXCR4 increases in-vivo glioma perivascular invasion, and reduces radiation induced apoptosis: A genetic knockdown study. Oncotarget. 2016, Dec 13;7 (50):83701–19.

[46] Griveau A, Seano G, Shelton SJ, Kupp R, Jahangiri A, Obernier K, et al. A Glial Signature and Wnt7 Signaling Regulate Glioma-Vascular Interactions and Tumor Microenvironment. Cancer Cell. 2018, May 14;33(5):874–89 e7.

[47] Lai CJ, Lin CY, Liao WY, Hour TC, Wang HD, Chuu CP, et al. CD44 Promotes Migration and Invasion of Docetaxel-Resistant Prostate Cancer Cells Likely via Induction of Hippo-Yap Signaling. Cells. 2019, Mar 30;8:4.

[48] Frentzas S, Simoneau E, Bridgeman VL, Vermeulen PB, Foo S, Kostaras, et al. Vessel co-option mediates resistance to anti-angiogenic therapy in liver metastases. Nat Med 2016, Nov;22(11):1294–302.

[49] Engbring JA, Kleinman HK. The basement membrane matrix in malignancy. J Pathol. 2003 Jul;200(4):465–70.

[50] Qin Y, Rodin S, Simonson OE, Hollande F. Laminins and cancer stem cells: partners in crime? Semin Cancer Biol 2017;45:3–12.

[51] Lugassy C, Eyden BP, Christensen L, Escande JP. Angio-tumoral complex in human malignant melanoma characterized by free laminin: ultrastructural and immunohistochemical observations. J Submicrosc Cytol Pathol 1997;29:19–28.

[52] Lugassy C, Dickersin GR, Christensen L, Karaoli T, LeCharpentier M, Escande JP, et al. Ultrastructural and immunohistochemical studies of the periendothelial matrix in human melanoma: evidence for an amorphous matrix containing laminin. J Cutan Pathol 1999;26(2):78–83.

[53] Giannelli G, Falk-Marzillier J, Schiraldi O, Stetler-Stevenson WG, Quaranta V. Induction of cell migration by matrix metalloprotease-2 cleavage of laminin-5. Science 1997;277(5323):225–8.

[54] Lugassy C, Torres-Muñoz JE, Kleinman HK, Ghanem G, Vernon SE, Barnhill RL. Over-expression of malignancy-associated laminins and laminin receptors by angiotropic human melanoma cells in a chick chorioallantoic membrane model. J Cutan Pathol 2009;36(12):1237–43.

[55] Oikawa Y, Hansson J, Sasaki T, Rousselle P, Domogatskaya A, Rodin S, et al. Melanoma cells produce multiple laminin isoforms and strongly migrate on α5 laminin(s) via several integrin receptors. Exp Cell Res 2011;317(8):1119–33.

[56] Carpenter PM, Sivadas P, Hua SS, Xiao C, Gutierrez AB, Ngo T, et al. Migration of breast cancer cell lines in response to pulmonary laminin 332. Cancer Med 2017;6(1):220–34.

[57] Tsubota Y, Ogawa T, Oyanagi J, Nagashima Y, Miyazaki K. Expression of laminin gamma2 chain monomer enhances invasive growth of human carcinoma cells in vivo. Int J Cancer 2010;127(9):2031–41.

[58] Teng Y, Wang Z, Ma L, Zhang L, Guo Y, Gu M, et al. Prognostic significance of circulating laminin gamma2 for early-stage non-small-cell lung cancer. Onco Targets Ther 2016;9:4151–62.

[59] Sun T, Patil R, Galstyan A, Klymyshyn D, Ding H, Chesnokova A, et al. Blockade of a laminin-411-notch axis with CRISPR/Cas9 or a nanobioconjugate inhibits glioblastoma growth through tumor-microenvironment cross-talk. Cancer Res 2019;79(6):1239–51.

[60] Hutchins EJ, Bronner ME. Draxin alters laminin organization during basement membrane remodeling to control cranial neural crest EMT. Dev Biol 2019;446(2):151–8.

[61] Prager BC, Xie Q, Bao S, Rich JN. Cancer stem cells: the architects of the tumor ecosystem. Cell Stem Cell 2019;24(1):41–53.

[62] Roth L, Kalev-Altman R, Monsonego-Ornan E, Sela-Donenfeld D. A new role of the membrane-type matrix metalloproteinase 16 (MMP16/MT3-MMP) in neural crest cell migration. Int J Dev Biol 2017;61(3-4-5):245–56.

[63] Virchow RLK, editor. Cellular pathology. 1859 special ed. London: John Churchill; 1978.

[64] Cofre J, Abdelhay E. Cancer is to embryology as mutation is to genetics: hypothesis of the cancer as embryological phenomenon. SciWorld J 2017;2017:3578090.

[65] Aiello NM, Stanger BZ. Echoes of the embryo: using the developmental biology toolkit to study cancer. Dis Model Mech 2016;9(2):105–14.

[66] Campbell K. Contribution of epithelial-mesenchymal transitions to organogenesis and cancer metastasis. Curr Opin Cell Biol 2018;55:30–5.

[67] Lee G, Hall III RR, Ahmed AU. Cancer stem cells: cellular plasticity, niche, and its clinical relevance. J Stem Cell Res Ther 2016;6(10) pii: 363.

[68] Saitoh M. Involvement of partial EMT in cancer progression. J Biochem 2018;164(4):257–64.

[69] Corselli M, Chen CW, Sun B, Yap S, Rubin JP, Péault B. The tunica adventitia of human arteries and veins as a source of mesenchymal stem cells. Stem Cells Dev 2012;21(8):1299–308.

[70] Burton GJ, Watson AL, Hempstock J, Skepper JN, Jauniaux E. Uterine glands provide histiotrophic nutrition for the human fetus during the first trimester of pregnancy. J Clin Endocrinol Metab 2002;87(6):2954–9.
[71] Burton GJ, Jauniaux E. Development of the human placenta and fetal heart: synergic or independent? Front Physiol 2018;9:373.
[72] Mitchell MJ, Denais C, Chan MF, Wang Z, Lammerding J, King MR. Lamin A/C deficiency reduces circulating tumor cell resistance to fluid shear stress. Am J Physiol Cell Physiol 2015;309(11):C736–46.
[73] Lugassy C. Concepts of cancer and metastasis: historical perspective. Melanoma: From Biology to Target. AACR; 2019.
[74] Paget S. The distribution of secondary growths in cancer of the breast. Lancet 1889;133(3421):571–3.
[75] Ribatti D, Mangialardi G, Vacca A. Stephen Paget and the 'seed and soil' theory of metastatic dissemination. Clin Exp Med 2006;6(4):145–9.
[76] Gandalovičová A, Vomastek T, Rosel D, Brábek J. Cell polarity signaling in the plasticity of cancer cell invasiveness. Oncotarget 2016;7(18):25022–49.
[77] Dupin E, Calloni GW, Coelho-Aguiar JM, Le Douarin NM. The issue of themultipotency of the neural crest cells. Dev Biol. 2018;444(Suppl 1):S47–59 Dec 1.
[78] George L, Dunkel H, Hunnicutt BJ, Filla M, Little C, Lansford R, et al. In vivo time-lapse imaging reveals extensive neural crest and endothelial cell interactions during neural crest migration and formation of the dorsal root andsympathetic ganglia. Dev Biol. 2016 May 1;413(1):70–85.
[79] Kulesa PM, Kasemeier-Kulesa JC, Teddy JM, Margaryan NV, Seftor EA, Seftor RE, et al. Reprogramming metastatic melanoma cells to assume a neural crest cell-like phenotype in an embryonic microenvironment. Proc Natl Acad Sci USA 2006;103 (10):3752–7.
[80] Bailey CM, Morrison JA, Kulesa PM. Melanoma revives an embryonic migration program to promote plasticity and invasion. Pigment Cell Melanoma Res 2012;25 (5):573–83.
[81] Tsai HH, Niu J, Munji R, Davalos D, Chang J, Zhang H, et al. Oligodendrocyte precursors migrate along vasculature in the developing nervous system. Science 2016;351(6271):379–84.
[82] Canoll P, Goldman JE. The interface between glial progenitors and gliomas. Acta Neuropathol 2008;116(5):465–77.
[83] Sakariassen PØ, Prestegarden L, Wang J, Skaftnesmo KO, Mahesparan R, Molthoff C, et al. Angiogenesis-independent tumor growth mediated by stem-like cancer cells. Proc Natl Acad Sci USA 2006;103(44):16466–71.
[84] Szabó A, Theveneau E, Turan M, Mayor R. Neural crest streaming as an emergent property of tissue interactions during morphogenesis. PLoS Comput Biol 2019;15(4): e1007002.
[85] McLennan R, Bailey CM, Schumacher LJ, Teddy JM, Morrison JA, Kasemeier-Kulesa JC, et al. DAN (NBL1) promotes collective neural crest migration by restraining uncontrolled invasion. J Cell Biol 2017;216(10):3339–54.
[86] Hadjimichael C, Chanoumidou K, Papadopoulou N, Arampatzi P, Papamatheakis J, Kretsovali A. Common stemness regulators of embryonic and cancer stem cells. World J Stem Cells 2015;7(9):1150–84.
[87] Hendrix MJ, Seftor EA, Seftor RE, Chao JT, Chien DS, Chu YW. Tumor cell vascular mimicry: novel targeting opportunity in melanoma. Pharmacol Ther 2016;159:83–92.
[88] Patel-Hett S, D'Amore PA. Signal transduction in vasculogenesis and developmental angiogenesis. Int J Dev Biol 2011;55(4–5):353–63.
[89] Burton GJ, Jauniaux E. Development of the human placenta and fetal heart: synergic or independent? Front Physiol 2018;9:373.

[90] Di Gregorio A, Bowling S, Rodriguez TA. Cell competition and its role in the regulation of cell fitness from development to cancer. Dev Cell 2016;38(6):621–34.
[91] Stine RR, Matunis EL. Stem cell competition: finding balance in the niche. Trends Cell Biol 2013;23(8):357–64.
[92] Mintz B, Illmensee K. Normal genetically mosaic mice produced from malignant teratocarcinoma cells. Proc Natl Acad Sci USA 1975;72(9):3585–9.
[93] Chen SH, Zhang BY, Zhou B, Zhu CZ, Sun LQ, Feng YJ. Perineural invasion of cancer: a complex crosstalk between cells and molecules in the perineural niche. Am J Cancer Res 2019;9(1):1–21.
[94] Friedl P, Alexander S. Cancer invasion and the microenvironment: plasticity and reciprocity. Cell 2011;147(5):992–1009.
[95] Erdogan B, Ao M, White LM, Means AL, Brewer BM, Yang L, et al. Cancer-associated fibroblasts promote directional cancer cell migration by aligning fibronectin. J Cell Biol 2017;216(11):3799–816.
[96] Ray A, Slama ZM, Morford RK, Madden SA, Provenzano PP. Enhanced directional migration of cancer stem cells in 3D aligned collagen matrices. Biophys J 2017;112 (5):1023–36.
[97] Friedl P, Gilmour D. Collective cell migration in morphogenesis, regeneration and cancer. Nat Rev Mol Cell Biol 2009;10(7):445–57.
[98] Doglioni G, Parik S, Fendt SM. Interactions in the (pre)metastatic niche support metastasis formation. Front Oncol 2019;9:219.
[99] Liu Q, Zhang H, Jiang X, Qian C, Liu Z, Luo D. Factors involved in cancer metastasis: a better understanding to "seed and soil" hypothesis. Mol Cancer 2017;16(1):176.
[100] Yeh AC, Ramaswamy S. Mechanisms of cancer cell dormancy – another hallmark of cancer? Cancer Res 2015;75(23):5014–22.
[101] Ingangi V, Minopoli M, Ragone C, Motti ML, Carriero MV. Role of microenvironment on the fate of disseminating cancer stem cells. Front Oncol 2019;9:82.
[102] Kienast Y, von Baumgarten L, Fuhrmann M, Klinkert WE, Goldbrunner R, Herms J, et al. Real-time imaging reveals the single steps of brain metastasis formation. Nat Med 2010, Jan;16(1):116–22.
[103] Clark AG, Vignjevic DM. Modes of cancer cell invasion and the role of the microenvironment. Curr Opin Cell Biol 2015;36:13–22.
[104] Kulesa PM, Teddy JM, Stark DA, Smith SE, McLennan R. Neural crest invasion is a spatially-ordered progression into the head with higher cell proliferation at the migratory front as revealed by the photoactivatable protein, KikGR. Dev Biol 2008;316(2):275–87.
[105] Li Y, Vieceli FM, Gonzalez WG, Li A, Tang W, Lois C, et al. In vivo quantitative imaging provides insights into trunk neural crest migration. Cell Rep 2019;26 (6):1489–500.
[106] Barbato L, Bocchetti M, Di Biase A, Regad T. Cancer Stem Cells and Targeting Strategies. Cells. 2019 Aug 18;8:8.
[107] Hindley CJ, Condurat AL, Menon V, Thomas R, Azmitia LM, Davis JA, et al. The Hippo pathway member YAP enhances human neural crest cell fate and migration. Sci Rep 2016;6:23208.
[108] Yan Y, Zuo X, Wei D. Concise review: emerging role of CD44 in cancer stem cells: a promising biomarker and therapeutic target. Stem Cells Transl Med 2015;4 (9):1033–43.
[109] Lai CJ, Lin CY, Liao WY, Hour TC, Wang HD, Chuu CP. CD44 promotes migration and invasion of docetaxel-resistant prostate cancer cells likely via induction of hippo-yap signaling. Cells 2019;8(4).

[110] Chang HY, Chang HM, Wu TJ, Chaing CY, Tzai TS, Cheng HL, et al. The role of Lutheran/basal cell adhesion molecule in human bladder carcinogenesis. J Biomed Sci 2017;24(1):61.

[111] Enomoto-Okawa Y, Maeda Y, Harashima N, Sugawara Y, Katagiri F, Hozumi K, et al. An anti-human lutheran glycoprotein phage antibody inhibits cell migration on laminin-511: epitope mapping of the antibody. PLoS One 2017;12(1):e0167860.

[112] Ridenour DA, McLennan R, Teddy JM, Semerad CL, Haug JS, Kulesa PM. The neural crest cell cycle is related to phases of migration in the head. Development 2014;141(5):1095–103.

[113] Fane ME, Chhabra Y, Smith AG, Sturm RA. BRN2, a POUerful driver of melanoma phenotype switching and metastasis. Pigment Cell Melanoma Res 2019;32(1):9–24.

[114] Maeda H, Khatami M. Analyses of repeated failures in cancer therapy for solid tumors: poor tumor-selective drug delivery, low therapeutic efficacy and unsustainable costs. Clin Transl Med 2018;7(1):11.

CHAPTER

5

Vasculogenic mimicry

Yun Cao[1] *and Chao-Nan Qian*[2]

[1]Department of Pathology, Sun Yat-sen University Cancer Center, Guangzhou, P.R. China, [2]Guangzhou Concord Cancer Center, Guangzhou, P.R. China

Introduction

Vasculogenesis refers to the creation of new blood vessels based on the absence of existing blood vessels, which typically involves the recruitment and differentiation of endothelial progenitor cells. In contrast, angiogenesis generates new blood vessels from budding in existing blood vessels. There is no angiogenesis in solid tumors, and the tumor size cannot exceed 2–3 cm in diameter [1]. Vasculogenic mimicry (VM) is a microvascular channel formed by invasive, metastatic, and hereditary dysregulated tumor cells. This process differs from angiogenesis, it occurs de novo in the absence of endothelial cells (tumor cells effectively mimic the true vascular endothelium). In 1999 it was first described by Maniotis et al. in uveal melanoma (Fig. 5.1) [2,3]. There are two main types of angiogenic mimicry: tubular and patterned. The former is similar in morphology to normal blood vessels, while the latter is consistent with the blood vessels, but it is significantly different from the former. After the vascular mimicry was proposed in 1999, vascular mimicry gradually progressed in renal cell carcinoma, breast cancer, ovarian cancer, primary gallbladder cancer, malignant esophageal squamous cell carcinoma, mesothelioma, alveolar rhabdomyosarcoma, and hepatocellular carcinoma [4]. The basement membrane stained with periodic acid Schiff (PAS) can be observed by light microscopy in these vascular ducts, and endothelial cells were not identified within these matrix-embedded channels by light microscopy, by transmission electron microscopy, or by using an immunohistochemical panel of endothelial cell markers [5].

Tumor Vascularization
DOI: https://doi.org/10.1016/B978-0-12-819494-2.00005-5

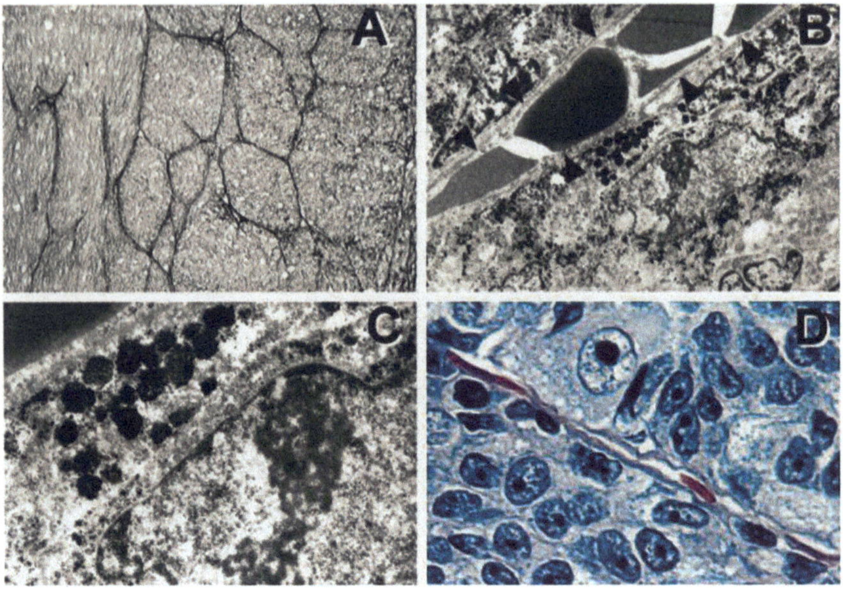

FIGURE 5.1 Transmission electron microscopy of melanoma microcirculation patterns. (A) Tissue section of primary uveal melanoma stained by PAS without hematoxylin counterstain. Areas from this tumor were microdissected and studied by transmission electron microscopy. (B) Scanning magnification transmission electron micrograph of a tubule containing a single-file column of red blood cells. This vascular channel is lined by a thin basal lamina (*arrowheads*) corresponding to the walls of the channel seen by conventional light microscopy. There are no endothelial cells lining the tubule. Tumor cells containing melanosomes and premelanosomes lie external to the basal lamina. (C) Higher magnification of this illustrating premelanosomes and melanosomes within the tumor. (D) Hematoxylin–eosin–stained tissue section of primary uveal melanoma. A vascular channel containing red blood cells is lined externally by tumor cells; endothelial cells are not identified. Original magnifications, 340 (A), 310,000 (B), 330,000 (C), and 3100 (D) [2].

Vasculogenic mimicry in prognosis and progress

The VM is associated with poor tumor differentiation and high malignancy. The density of VM in poorly differentiated tumors is higher than that of well-differentiated tumors. At the same time, tumor infiltration and metastasis are also related to VM. Groups with high VM density are more likely to infiltrate and metastasis than the control group. In hepatocellular carcinoma (HCC), the relationship of VM and the Edmondson grade has been discussed. The positive rate of VM in Edmondson grade III and grade IV tumors (higher grade) was statistically higher than that of Edmondson grade I and grade II tumors (lower grade) [6,7]. The presence of VM in prostate cancer tissues is associated with higher Gleason scores, tumor, node, and metastasis (TNM) staging, lymph node, and distant metastasis [8,9]. Similar results were found in gastric cancer. The VM in early gastric carcinoma (GC) tissues was associated with tumor size, differentiation, depth of tumor invasion, stage, lymph node metastasis, and tumor emboli in microvessels [9]. In a recent study, a causal relationship between VM and metastasis is proposed. In this study using breast cancer animal models, the metastatic populations within a heterogenous tumor demonstrate their ability of forming VM to ensure other metastatic cells travel into circulation. The two identified proteins secreted by metastatic tumor cells, SERPINE2 and secretory leukocyte protease inhibitor, are responsible for promoting VM formation. However, the tumor cells forming vascular network are morphologically altered to be thin layer cells similar to normal endothelial cells. There is no evidence to show that these cells could be transformed back to active metastatic tumor cells. Therefore it is more reasonable to speculate that the metastatic cancer cell populations have a potential to form VM by sacrificing a small portion within the population and ensuring metastasis of the remaining cells; the cells forming VM are the ones most likely undergoing differentiation. This orchestrating action is partially supported by the evidence that in metastasis of pancreatic cancer cells, multiple clones of cancer cells occur in different phases of metastasis, indicating the heterotypic interactions between tumor subpopulations contributing to metastasis progression [10]. VM is associated with poor prognosis of multiple solid treatments, reduced survival, and high risk of cancer recurrence [11]. The appearance of VM also means that tumor cells may be insensitive to traditional radiotherapy and chemotherapy. Conventional chemotherapy in Merkel cell carcinoma (MCC) is difficult to cure in the areas that rich in vascular mimicry. That in etoposide and carboplatin-resistant MCC cells, two of the three laminin isoforms showed a statistically significant increase in mRNA levels, which has a positively correlated with VM, have also been found in MCC [12].

Molecular pathway of vasculogenic mimicry

Various molecular pathway has been reported that may involve in VM by many articles. However, at present the precise molecular mechanisms involved in VM are still unknown. This part discusses the important function of some molecular in the progress of VM formation [13].

Hypoxia/hypoxia-inducible factor pathway

Tumor cells prefer to grow using glycolysis rather than oxidative phosphorylation. This metabolic approach is suitable to hypoxic conditions, making cancer cells more resistant to hypoxia. Hypoxia has been recognized as a primary physiological regulator of the angiogenic switch [1]. Under hypoxic conditions, a pathway involving a key oxygen response regulator called hypoxia-inducible factor (HIF) is turned on. The alpha subunit of HIF-1 (HIF-1α) is a nuclear factor commonly found in mammals and is a recognized mediator of cancer response to hypoxia. HIF-1α is degraded shortly after expression in the cytoplasm under normoxic conditions. However, under hypoxic conditions, HIF-1α protein can be transferred to the nucleus to bind to the β subunit of HIF-1 to form HIF-1 heterodimer. Studies have shown that HIF-1α is related to angiogenesis. The core recognition sequence of the promoter sequence of HIF-1 hypoxia-inducible gene combines 5′-RCGTG-3′ to bind and promote transcription and translation of these genes [14]. Sun et al. have found that hypoxia could induce more VM channel formation and elevated HIF-1α expression in highly aggressive gallbladder carcinoma (GBC). However, HIF-1α siRNA efficiently knocked down HIF-1α expression and GBC VM networks under either normoxic or hypoxic conditions. It means HIF may also play an important role in VM formation [4]. In HCC, the similar result has been demonstrated that there exists an important regulatory axis in tumor hypoxic microenvironment involving elevated levels of LOXL2 induced by HIF-1α. This increase results in the subsequent promotion of VM [15]. In vivo data showed that the human colorectal carcinoma cell line HCT-116 cells formed vascular channels. In addition, the formation of these channels was associated with increased expression of HIF-1α. Moreover, knockdown of HIF-1α in HCT-116 tumor xenografts prevented VM formation and reduced gene expression changes implicated in the process of epithelial-mesenchymal transition (EMT) which process is similar to that of VM (Fig. 5.2) [16].

Vascular endothelial–cadherin pathway

One of the important vascular-associated molecule shown to be involved in VM is vascular endothelial (VE)–cadherin (CDH5). VE–cadherin is a transmembrane glycoprotein of the cadherin family that promotes homotypic cell–cell interactions and is considered specific for vascular endothelia and critical for vasculogenic events [17]. VE–cadherin was exclusively expressed by highly aggressive melanoma cells and was undetectable in the poorly aggressive tumor cells. Down-regulated VE–cadherin expression in the aggressive melanoma

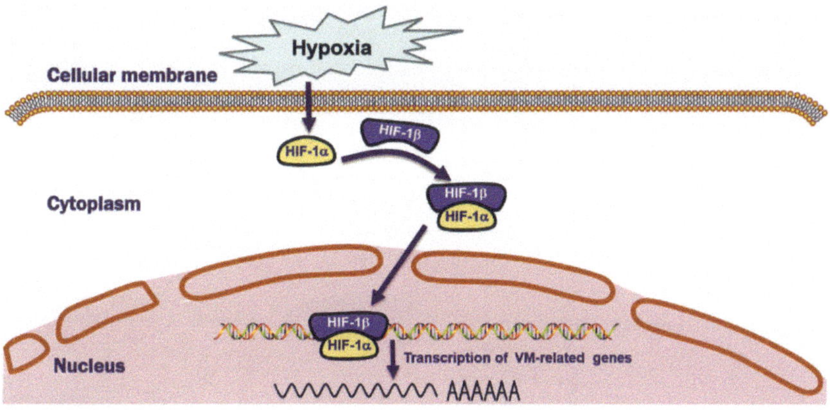

FIGURE 5.2 Hypoxia induced the expression of VM-related genes in tumor cells.

cells and their ability to form vasculogenic networks immediately destroyed. It directly tested the hypothesis that VE–cadherin expressed by aggressive melanoma tumor cells resulted in their ability to mimic endothelial cells by forming patterned vasculogenic networks [18]. Other molecules that have been reported to play a role in E-cadherin pathway involved in VM formation. HMGA2 and Twist1 are associated with VM in GC; moreover, they observed a decrease in VE–cadherin following Twist1 knockdown in cells overexpressing HMGA2. This study indicates that HMGA2 promotes VM in GC via Twist1–VE–cadherin signaling and influences the prognosis of patients with GC [19]. The possible mechanism is that elevated Twist1 can bind to the VE–cadherin promoter and enhances its activity in a transactivation assay. These similar results also demonstrated in human PCa PC-3 cells [20]. In nonsmall cell lung cancer (NSCLC) cells investigated that netrin-1 is involved in VM formation. Overexpression of netrin-1 promoted VM development in contrast to the vector-treated group, concomitant with an increase in the expression of VE–cadherin. Consistently, netrin-1 inhibition also antagonized the expression of VE–cadherin. This result suggests that netrin-1 promote VM in NSCLC may through VE–cadherin [21]. EphA2 is a member of the Eph family of tyrosine receptor kinase expressed (RTK) on the cell membrane [22]. VE–cadherin binds to EphA2 then phosphorylate EphA2 [18,23]. Phosphorylated EphA2 phosphorylates and activates downstream FAK and PI3K [24,25]. Activated FAK continues phosphorylates and activates ERK1/2 [24]. Active PI3K and ERK1/2 both can regulate the transition of pro-MT1-MMP to active MT-MMP, which subsequently activates pro-MMP2 [26,27]. MT1-MMP and MMP-2 promote laminin5γ chain splits into fragments $\gamma 2'$ and γ2x [28], which play an important role in tumor invasion, metastasis, and VM (Fig. 5.3) [29].

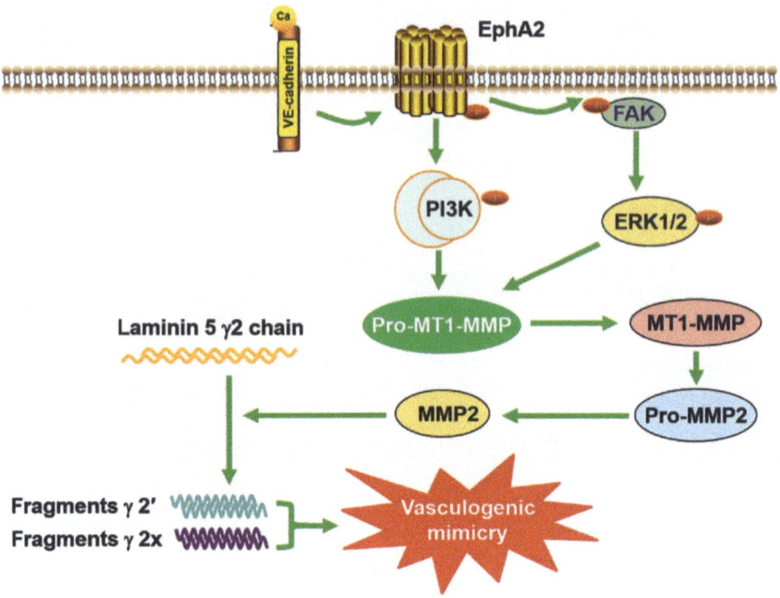

FIGURE 5.3 Signaling pathways of VE–cadherin–induced VM in tumor cells.

Long noncoding RNAs pathway

Long noncoding RNAs (lncRNAs) upregulated expression which had been proved as classical markers of VM. Aberrant expression of some lncRNAs has been reported in various tumors, and lots of lncRNAs played important roles in regulating biological processes of various tumors [30]. It suggesting that lncRNAs might be involved in the regulation of tumor progression [31]. That the expression levels of lncRNAs in tumor tissues were positively correlated with tumor VM formation also be demonstrated [32–35]. That knockdown of lncRNAs inhibited VM formation. This evidence suggested that aberrant expression of lncRNAs might play a pivotal role in the VM formation of tumor [36]. lncRNAs are a kind of noncoding RNA that is longer than 200 nucleotides and does not possess protein coding ability [30]. The average VM density in tumor tissues formed in lncRNA LINC00312 overexpression cells was significantly higher than that in control cells in lung adenocarcinoma. In accordance, knockdown of LINC00312 reduced the ability of migration and invasion. LINC00312 induces VM of ADC cell lines through YBX1. A pro-oncogenic role for YBX1 is suggested by its ability to promote migration and invasion of tumor cells as angiogenesis switch. When YBX1 was knockdown, LINC00312-mediated VM was blocked [37]. In glioma, the expression of lncRNA LINC00339 and HOXA-AS2 was upregulated and positively correlated with glioma VM formation. Knockdown of LINC00339 and HOXA-AS2 inhibited glioma cell proliferation, migration, invasion, and VM formation, meanwhile downregulating the expression of VM-related molecular matrix metalloproteinase (MMP-2, MMP-9, and MMP-14) [33,35]. MALAT1 is an oncogenic lncRNA that has been found to promote the proliferation of many malignant cell types and nonmalignant human umbilical vein endothelial cells. That MALAT1 expression was tightly associated with densities of VM and endothelial vessels. MALAT1 knockdown markedly reduced GC cell migration, invasion, tumorigenicity, metastasis, and VM formation [34]. MALAT1 also palys a similar role in NSCLC VM formation promoted by ERβ via altering the ERβ/MALAT1/miR145-5p/NEDD9 signaling [31]. lncRNA TP73 antisense RNA 1 (TP73-AS1) was upregulated in VM positive triple negative breast cancer (TNBC) tissues. Knockdown of TP73-AS1 suppressed TNBC cell line MDA-MB-231 cell VM formation in vitro [36]. The lncRNA n339260 level correlated with HCC VM in an animal model has also proposed by Zhao et al. [32]. A determinate mechanism about lncRNAs contribute to VM formation is still unknown. lncRNAs can induce cancer stem cells—like phenotype that promote tumors to mimic vascular endothelial cells to form tube [38].

Vascluar endothelial growth factor pathway

The vascular endothelial growth factor (VEGF) gene family includes four members in mice and humans (VEGF-A, -B, -C, and -D) and a related protein, placental growth factor. Signals from VEGF pathways control many biological effects, the best studied being effects on endothelial cell migration, proliferation, and cell survival, and the expression of downstream genes in endothelial cells and tumor cells [1]. VEGF-A is one of the family of five angiogenic growth factors that plays a crucial role in tumor angiogenesis by recruiting and stimulating the proliferation of endothelial cells in a vascular regions of rapidly growing tumors [39]. Moreover, VEGF expression does not increase and sometimes even drops when VM exists, even though tumor aggressively increases. Thus whether VEGF has role in VM formation may not so clear [40]. The process VEGF-a → EphA2 → MMPs → VM is the main pathway for VM formation, and VEGF-a appears to play an important role in the formation of VM based on their in vitro assays and clinical immunohistochemical analyses [39,41]. In prostate cancer that VEGF effectively promoted the formation of capillary-like structures, and highlighted that VEGF and the phosphoinositide 3-kinase/AKT pathway exert a positive feedback regulation in the process of VM. While article point out that the existence of VM might help tumor cells obtain enough blood and sufficient oxygen supply, leading to a decrease in the secretion of the VEGF inducer and a consequent drop in VEGF expression in tumor cells. These results point to a causal relationship between VM and VEGF [40].

Nodal pathway

Nodal is an embryonic morphogen, which belongs to the transforming growth factor superfamily. The molecules related to EMT increase in the Nodal overexpression cell lines than the knockdown B16 melanoma stable cell lines. Snail and Slug are crucial for Nodal-induced EMT and can drive the progress of EMT. And Nodal upregulates Snail and Slug partly through ALK/Smads and PI3k/AKT pathways [42]. The similar pathway is also reported in melaoma. The Nodal precursor secreted by tumor cells is cleaved and activated by proprotein convertase (SPC), Pace-4, and Furin. After binding with heterodimeric complexes including type I and type II activin-like kinase receptors, Nodal phosphorylates Smad-2/3 and then binds with Smad-4, translocates to the nucleus to regulate downstream gene expression. Then nodal is a biomarker of tumor, Nodal is only express in invasive vertical growth phase and metastatic melanoma lesions [42,43]. Meanwhile, nodal is vital in breast carcinoma to sustain continued growth and plasticity [44]. As mentioned, Nodal signaling is known to support VM and is associated with expression of VE–cadherin [42].

Therapeutic strategy for vasculogenic mimicry

VM has also been reported to be resistant to endostatin and angiogenesis inhibitors such as TPN-470 in melanoma tumor cells and the B16F10 mouse melanoma model [45]. Therefore therapies that target angiogenesis should not be the only strategies that target the tumor microcirculation. Additional tumor-targeted therapies for nonangiogenic pathways of tumor perfusion and metastasis need to be considered.

Targeting Hypoxia/hypoxia-inducible factor pathway

As mentioned earlier, Hypoxia/HIF pathway palys an important role in angiogenesis and VM. HIF-targeted agent, 2-Methoxyestradiol, may be warranted for inhibiting formation of VM [46]. Since mTOR regulates HIF-1α expression and its transcriptional activity, rapamycin was proposed as a promising hypoxia-related therapeutic approach in melanoma treatment [47]. Myo-inositol trispyrophosphate (ITPP) is an effector of membrane-permeable allosteric hemoglobin, which enhances the oxygen release capacity of red blood cells (RBCs), thereby increasing the oxygen content in the microenvironment and reducing the hypoxia effect. In vitro experiments have shown that ITPP-loaded RBCs have low O_2 affinity, thus significantly reducing the production of HIF-1α and VEGF-A by cocultured endothelial cells and cells to prevent the formation of vascular networks. An in vivo animal model of rat hepatocellular carcinoma further demonstrates that ITPP has a high tumor growth–stable growth, allowing long-term survival and even curing most tumors [1].

Targeting vascular endothelial–cadherin pathway

Knocking out the EphA2 gene, downregulating VE–cadherin, targeting MMPs and the Laminin-5γ2 chain, and inhibiting the PI3K pathway are potential target strategies for inhibition of VM [38]. Genistein was found to suppress VM by inhibiting VE-cadherin expression [48], while chemically modified tetracycline-3 can simultaneously inhibit the expression of VE–cadherin, MMP-2, and MT1-MMP [49]. The capacity of inhibition of PTK to decrease VM could be one of the reasons why such drugs could be successfully used in the treatment of gastrointestinal stromal tumors (GIST) [50]. Doxycycline inhibits MMP-2 expression [51].

Targeting vascular endothelial growth factor pathway

VEGF can be a strong target to block both endothelial and nonendothelial tumor angiogenesis. Inhibition of VEGF resulted in the disability of osteosarcoma cells to completely form an angiogenesis-like network in three-dimensional culture [5]. Thalidomide influences VM channel formation in melanoma by inhibiting MMP-2 and VEGF expression [52]. The FDA-approved antiangiogenic 1st agent is Bevacizumab, a humanized anti-VEGF monoclonal antibody, in combination with cytotoxic chemotherapy or immunotherapy for metastatic colorectal cancer, nonsquamous epithelium. It has benefited NSCLC and renal cell carcinoma.

Targeting other pathway

Recent reports have revealed that human high-grade invasive tumor that expressed high levels of COX-2 proteins had detectable vascular channels. There was little evidence of VM formation in low-grade tumors with low COX-2 expression. Celecoxib, a COX-2 inhibitor, may block vascular channel formation, but addition of prostaglandin E2 (PGE2) abrogates these inhibitory effects [53].

Conclusion

VM refers to the ability that the highly aggressive tumors to form blood vessels by tumor cells instead of endothelial cells, which has been found in various kinds of malignant tumors. VM is more easy to find in poorly differentiated tumors than in well-differentiated tumors. VM can serve as a significant marker for tumor metastasis, a poor prognosis, worse survival, and the highest risk of cancer recurrence for patients in various tumors. Various molecular pathways have been reported in many articles. The molecules (Hypoxia/HIF pathway, VEGF pathway, and Nodal pathway) may play an important role in the formation of VM. However, the detailed mechanism remains unclear and requires more and further research. Even so, the targeting for Hypoxia/HIF pathway, VE-cadherin pathway, VEGF pathway, etc. has been proved to be useful for inhibition of VM. Meanwhile VM present in tumors predicts that the traditional radiotherapy and chemotherapy may not be so effective, thus anti-VM therapy may be important strategy in tumor therapy in the future.

References

[1] Qin L, Bromberg-White JL, Qian CN. Opportunities and challenges in tumor angiogenesis research: back and forth between bench and bed. Adv Cancer Res 2012;113:191–239.
[2] Maniotis AJ, et al. Vascular channel formation by human melanoma cells in vivo and in vitro: vasculogenic mimicry. Am J Pathol 1999;155(3):739–52.
[3] Folberg R, Hendrix MJ, Maniotis AJ. Vasculogenic mimicry and tumor angiogenesis. Am J Pathol 2000;156(2):361–81.
[4] Sun W, et al. Overexpression of HIF-1α in primary gallbladder carcinoma and its relation to vasculogenic mimicry and unfavourable prognosis. Oncol Rep 2012;27(6):1990–2002.
[5] Yinsheng C, et al. Vasculogenic mimicry-potential target for glioblastoma therapy: an in vitro and in vivo study. Med Oncol 2012;29(1):324–31.
[6] Sun T, et al. Expression and functional significance of Twist1 in hepatocellular carcinoma: its role in vasculogenic mimicry. Hepatology 2010;51(2):545–56.
[7] Baocun S, et al. Vasculogenic mimicry is associated with high tumor grade, invasion and metastasis, and short survival in patients with hepatocellular carcinoma. Oncol Rep 2006;16(4):693–8.
[8] Ahmadi SA, et al. Practical application of angiogenesis and vasculogenic mimicry in prostatic adenocarcinoma. Arch Iran Med 2010;13(6):498–503.
[9] You X, et al. Prognostic significance of galectin-1 and vasculogenic mimicry in patients with gastric cancer. Onco Targets Ther 2018;11:3237–44.

[10] Qian CN, et al. Revisiting tumor angiogenesis: vesselco-option, vessel remodeling, and cancercell-derived vasculature formation. Chin J Cancer 2016;35(1):1–6.
[11] Li S, et al. The hypoxia-related signaling pathways of vasculogenic mimicry in tumor treatment. Biomed Pharmacother 2016;80:127–35.
[12] Cecilia L, et al. Merkel cell carcinoma expresses vasculogenic mimicry: demonstration in patients and experimental manipulation in xenografts. Lab Invest 2014;94(10):1092–102.
[13] Paulis YW, Verheul HM, Griffioen AW. Signalling pathways in vasculogenic mimicry. Biochim Biophys Acta 2010;1806(1):18–28.
[14] Xiang T, et al. Vasculogenic mimicry formation in EBV-associated epithelial malignancies. Nat Commun 2018;9(1):5009.
[15] Wang M, et al. HIF-1α promoted vasculogenic mimicry formation in hepatocellular carcinoma through LOXL2 up-regulation in hypoxic tumor microenvironment. J Exp Clin Cancer Res 2017;36(1):60.
[16] Li W, et al. Hypoxia-induced vasculogenic mimicry formation in human colorectal cancer cells: involvement of HIF-1a, Claudin-4, and E-cadherin and Vimentin. Sci Rep 2016;6:37534.
[17] Kirschmann DA, et al. Molecular pathways: vasculogenic mimicry in tumor cells: diagnostic and therapeutic implications. Clin Cancer Res 2012;18(10):2726–32.
[18] Hendrix MJ, et al. Expression and functional significance of VE-cadherin in aggressive human melanoma cells: role in vasculogenic mimicry. Proc Natl Acad Sci USA 2001;98(14):8018–23.
[19] Sun J, et al. HMGA2 promotes vasculogenic mimicry and tumor aggressiveness by upregulating Twist1 in gastric carcinoma. Sci Rep 2017;7(1):2229.
[20] Yeo C, Lee HJ, Lee EO. Serum promotes vasculogenic mimicry through the EphA2/VE-cadherin/AKT pathway in PC-3 human prostate cancer cells. Life Sci 2019;221:267–73.
[21] Zhang X, et al. Netrin-1 elicits metastatic potential of non-small cell lung carcinoma cell by enhancing cell invasion, migration and vasculogenic mimicry via EMT induction. Cancer Gene Ther 2017;25(1–2):1.
[22] Margaryan NV, et al. EphA2 as a promoter of melanoma tumorigenicity. Cancer Biol Ther 2009;8(3):279–88.
[23] Qiao L, et al. Advanced research on vasculogenic mimicry in cancer. J Cell Mol Med 2015;19(2):315–26.
[24] Duxbury MS, et al. Ligation of EphA2 by Ephrin A1-Fc inhibits pancreatic adenocarcinoma cellular invasiveness. Biochem Biophys Res Commun 2004;320(4):1096–102.
[25] Hess AR, et al. Molecular regulation of tumor cell vasculogenic mimicry by tyrosine phosphorylation: role of epithelial cell kinase (Eck/EphA2). Cancer Res 2001;61(8):3250.
[26] Hess AR, et al. Focal adhesion kinase promotes the aggressive melanoma phenotype. Cancer Res 2005;65(21):9851–60.
[27] Hess AR, et al. Phosphoinositide 3-kinase regulates membrane Type 1-matrix metalloproteinase (MMP) and MMP-2 activity during melanoma cell vasculogenic mimicry. Cancer Res 2003;63(16):4757.
[28] Koshikawa N, et al. Role of cell surface metalloprotease Mt1-Mmp in epithelial cell migration over laminin-5. J Cell Biol 2000;148(3):615–24.
[29] Seftor RE, et al. Targeting the tumor microenvironment with chemically modified tetracyclines: inhibition of laminin 5 gamma2 chain promigratory fragments and vasculogenic mimicry. Mol Cancer Ther 2002;1(13):1173–9.
[30] Zhang L, et al. Long non-coding RNAs in Oral squamous cell carcinoma: biologic function, mechanisms and clinical implications. Mol Cancer 2019.
[31] Yu W, et al. Estrogen receptor β promotes the vasculogenic mimicry (VM) and cell invasion via altering the lncRNA-MALAT1/miR-145-5p/NEDD9 signals in lung cancer. Oncogene 2019;38(8):1225–38.

[32] Zhao X, et al. The long non-coding RNA n339260 promotes vasculogenic mimicry and cancer stem cell development in hepatocellular carcinoma. Cancer Sci 2018;109 (Spec No. 1).
[33] Guo J, et al. Long non-coding RNA LINC00339 stimulates glioma vasculogenic mimicry formation by regulating the miR-539-5p/TWIST1/MMPs axis. Mol Ther Nucleic Acids 2018;10:170–86.
[34] Li Y, et al. Long non-coding RNA MALAT1 promotes gastric cancer tumorigenicity and metastasis by regulating vasculogenic mimicry and angiogenesis. Cancer Lett 2017;395:31.
[35] Gao Y, et al. Long non-coding RNA HOXA-AS2 regulates malignant glioma behaviors and vasculogenic mimicry formation via the MiR-373/EGFR axis. Cell Physiol Biochem 2017;45(1):131–47.
[36] Tao W, et al. Knockdown of long non-coding RNA TP73-AS1 suppresses triple negative breast cancer cell vasculogenic mimicry by targeting miR-490-3p/TWIST1 axis. Biochem Biophys Res Commun 2018;504(4):629–34.
[37] Peng Z, et al. The long noncoding RNA LINC00312 induces lung adenocarcinoma migration and vasculogenic mimicry through directly binding YBX1. Mol Cancer 2018;17(1):167.
[38] Chen YS, Chen ZP. Vasculogenic mimicry:a novel target for glioma therapy. Chin J Cancer 2014;33(2):74–9.
[39] Wang J, et al. Functional significance of VEGF-a in human ovarian carcinoma: role in vasculogenic mimicry. Cancer Biol Ther 2008;7(5):758–66.
[40] Zhang S, Zhang D, Sun B. Vasculogenic mimicry: current status and future prospects. Cancer Lett 2007;254(2):157–64.
[41] Schnegg CI, et al. Induction of vasculogenic mimicry overrides VEGF-A silencing and enriches stem-like cancer cells in melanoma. Cancer Res 2015;75(8):1682.
[42] Topczewska JM, et al. Embryonic and tumorigenic pathways converge via Nodal signaling: role in melanoma aggressiveness. Nat Med 2006;12(8):925–32.
[43] Lynne-Marie P, et al. Role of nodal signaling and the microenvironment underlying melanoma plasticity. Pigment Cell Melanoma Res 2010;21(3):348–57.
[44] Strizzi L, et al. Emerging roles of nodal and Cripto-1: from embryogenesis to breast cancer progression. Breast Dis 2008;29(1):91.
[45] Fan YZ, et al. Molecular regulation of vasculogenic mimicry in tumors and potential tumor-target therapy. World J Gastrointest Surg 2010;2(4):117–27.
[46] Mooberry SL. Mechanism of action of 2-methoxyestradiol: new developments. Drug Resist Updat 2003;6(6):355–61.
[47] Michaylira CZ, Hiroshi N. Hypoxic microenvironment as a cradle for melanoma development and progression. Cancer Biol Ther 2006;5(5):476–9.
[48] Cong R, et al. Effect of genistein on vasculogenic mimicry formation by human uveal melanoma cells. J Exp Clin Cancer Res 2009;28(1):1–7.
[49] Lokeshwar BL, et al. Inhibition of cell proliferation, invasion, tumor growth and metastasis by an oral non-antimicrobial tetracycline analog (COL-3) in a metastatic prostate cancer model. Int J Cancer 2010;98(2):297–309.
[50] Zhao H, Gu XN. Study on vasculogenic mimicry in malignant esophageal stromal tumors. World J Gastroenterol 2008;14(15):2430–3.
[51] Sun B, et al. Doxycycline influences microcirculation patterns in B16 melanoma. Exp Biol Med 2007;232(10):1300.
[52] Zhang S, et al. Thalidomide influences growth and vasculogenic mimicry channel formation in melanoma. J Exp Clin Cancer Res 2008;27(1):60.
[53] Basu GD, et al. A novel role for cyclooxygenase-2 in regulating vascular channel formation by human breast cancer cells. Breast Cancer Res 2006;8(6):R69.

CHAPTER

6

Postnatal vasculogenesis

Laetitia Andrique[1,2], Gaelle Recher[3,4], Pierre Nassoy[3,4] and Andréas Bikfalvi[1,2]

[1]LAMC, Laboratoire de l'Angiogenèse et du Microenvironnement des Cancers (Inserm U1029), Pessac, France, [2]Université de Bordeaux, Pessac, France, [3]LP2N, Laboratoire Photonique Numérique et Nanosciences, Univ. Bordeaux, Talence, France, [4]Institut d'Optique Graduate School & CNRS UMR 5298, Talence, France

Introduction

During embryogenesis, the vascular network is formed by two distinct mechanisms, vasculogenesis and angiogenesis [1,2]. In adult, neovascularization is thought to arise only from angiogenesis, while vasculogenesis is restricted to embryonic life. This widely accepted view was challenged when endothelial progenitor cells (EPCs) were found in the blood circulation in adults [3,4].

At this stage, it is important to clarify the terminology used in this chapter. First, angiogenesis is defined as the formation of blood vessels and capillaries by the expansion of preexisting ones, while vasculogenesis classically requires the differentiation of a precursor cell and its incorporation into nascent vessels. Postnatal vasculogenesis (PNV) is a term that now refers to the formation of new blood vessels after birth and during adult life. It is thought to originate from EPCs that are recruited at sites where new vessel formation is required.

Another aspect closely related to PNV is arteriogenesis (and may be, in fact, considered as a form of PNV). In this case, vessel expansion by the proliferation of vascular cells inside the vessel wall occurs. EPCs may be incorporated into the arterial vessel wall, but the principal mechanisms of arteriogenesis are (1) expansion by proliferation of

Tumor Vascularization
DOI: https://doi.org/10.1016/B978-0-12-819494-2.00006-7

preexisting vascular cells within the vessel wall or (2) assembly of vascular cells that have migrated from vascular sprouts [5,6].

We will also provide a critical discussion about the importance of PNV especially related to the role of EPCs in physiology and pathology since the importance of their role has been challenged [7]. PNV encompasses also an additional aspect. Vascular tissue engineering represents, in fact, a form of artificial PNV, and this will also be discussed.

We are focusing in this chapter solely on blood vascular cells and we are not discussing lymphatic vasculogenesis, which is certainly an interesting topic and should be part of another article written by specialists in the field.

Vasculogenesis and angiogenesis

The circulatory system is an organ system which rapidly develops during embryonic life. It transports blood cells, nutrients, oxygen, CO_2, waste material, and cytokines to and from targeted tissues. Prenatal vasculogenesis is the neoformation of blood vessel from progenitor cells whereas angiogenesis constitutes an amplification mechanism of the preexisting vasculature in both, embryonic and adult life.

For a long time, angiogenesis was used as a generic term for vascularization but it does not apply in all cases and, thus, a clarification is needed.

First, prenatal vasculogenesis appears during embryonic life and represents the de novo formation of blood vessels from the differentiation of mesodermal cells. This differentiation process takes place in blood islands which first appears in mice at approximately days 13–15 of gestation, and consists of a loose inner mass of embryonic hematopoietic stem cells and an outer luminal layer of angioblasts/EPCs. This first phase of vasculogenesis is initiated by a combination of bone marrow (BM), Wnt, and Notch signaling pathways. Essential steps in this process are: (1) the birth of angioblasts; (2) angioblasts aggregation; (3) elongation of angioblasts into cord-like structures; (4) the organization of isolated vascular segments into capillary-like-networks and concomitant with step 4; and (5) endothelialization and humanization [8]. EPCs and hemangioblasts present in blood islands share antigens like Flk-1, Tie-2, Sca-1, and CD34, and will differentiate, respectively, into endothelial cells (ECs) and hematopoietic cells.

The succession of steps is well organized, synchronized during embryogenesis and, most of all, highly regulated by specific factors such as vascular endothelial growth factor (VEGF) and Ang1 [9]. Indeed, mice deficient for Ang1 rapidly die at early embryonic stage due to defects in vessel formation, while a deletion of VEGF-R2 or Ang1 receptor (Tie-2) is also embryonic lethal [10,11].

In the embryo, after the initial phase of vasculogenesis, angiogenesis begins. The primary vascular network is expanded under the control of the VEGF-A signaling, guidance cues, and arterial/venous specification mechanisms. Angiogenesis appears in both the embryo and the adult life by sprouting (derivation and growth from tip cells) or intussusception (splitting) of blood vessels [12].

Angiogenesis is implicated in both physiological (i.e., placenta formation, menstrual cycles) and pathological (chronic inflammation, neoplasia, cancer) processes. Nevertheless, both vasculogenesis and angiogenesis require EC proliferation, migration, and 3D organization into a tubular tissue.

Vasculogenesis, which was for long time thought to be restricted to embryos, was found also to occur in the adult life and is therefore designated as PNV. In 1997–1999 Asahara and collaborators reported the presence of CD34 + EPCs in the adult blood circulation. They also showed that heterologous, homologous, and autologous systemic transplantation of these CD34 + EPCs isolated from circulating blood, resulted in incorporation into sites of neovascularization. This work raised controversies regarding the origin of these progenitor cells, the choice of surface markers for their isolation, and the choice of the isolation/and amplification techniques. Ten years after, the contribution of EPCs to vessel development has been challenged by Purhonen et al. [7] who did not attribute a major role of EPCs in postnatal vascular development. This debate is still not settled and will be discussed later in this book (See Section Critical assessment of the importance of PNV versus angiogenesis).

Another form of vasculogenesis is arteriogenesis. Arteriogenesis is classically defined as the enlargement of preexisting blood vessels in response to physical or molecular stimuli. However, this definition, in the light of more recent research, is too restrictive and should be broadened. We should refer as arteriogenesis as an enlargement or assembly of arterial vessels initiated by ECs.

Postnatal vasculogenesis in physiological and pathological conditions

Endothelial progenitor cells

Stem cells are basically characterized by their self-renewal and clonogenicity capacities, while progenitor cells differ by the lack of self-renewal capacity. In the case of EPCs, they are unique because they keep the capacity of self-renewability, clonogenicity, and

TABLE 6.1 Expression markers between hemangioblasts, EPCs, and differentiated EC [13–15].

Hemangioblast	EPC	ECs
CD34 + CD133 + VEGFR-2 + / −	CD34 + CD133 + VEGFR-2 + / − CD45 + CD31 + / − CD177 + VE-Cadherin + / − vWf + / −	CD34 + CD133 − VEGFR-2 + CD45 − CD31 + CD177 − VE-Cadherin + vWf +

EC, endothelial cell; *EPC*, endothelial progenitor cell; *vWf*, von Willabrand factor.

differentiation [13]. EPCs are derived from hemangioblasts and differ from ECs by specific cell surface expression marker and by the expression level of these markers (Table 6.1).

Several cell surface markers are currently used in order to identify EPC such as VEGFR-2, CD34 (hematopoietic stem cell marker), and CD133 (hematopoietic stem and progenitor marker). Although it is difficult to perfectly distinguish EPCs from ECs, expression level of markers such as CD31, von Willebrand factor, and VE-cadherin (vascular endothelial cadherin) is used to identify cells in a more advanced stage of endothelial maturation. Moreover, the loss of progenitor markers such as CD177, CD133, and CD45 was observed in differentiated ECs.

Different steps in postnatal vasculogenesis

The majority of human EPCs (hEPCs) are located within a stem cell niche in BM and can be found in peripheral blood where it represents 0.0001%–0.01% of the total cells [16]. In BM, these cells are in a quiescent state and surrounded by stromal cells in a microenvironment characterized by low oxygen tension and high levels of chemoattractant molecules [17]. In case of vascular injury, PNV is initiated by the release of hEPCs from the BM to the circulation (Step 1, Fig. 6.1). This first step is followed by stem cell migration via the circulatory system to the injury zone (homing). How these cells reach the site of injury is not completely understood but it has been described that they can be guided by the concentration gradient of different chemoattractant molecules [18]. For example, it has been described that hEPC can invade and migrate (Step 2) under the influence of angiogenic growth factors such as VEGF-A, VEGF-B, stroma cell-derived factor 1, and insulin-like growth factor-I [19]. When arrived at targeted sites, hEPCs have the capacity to: (1) invade inside the tissue (Step 3) and (2) then differentiate

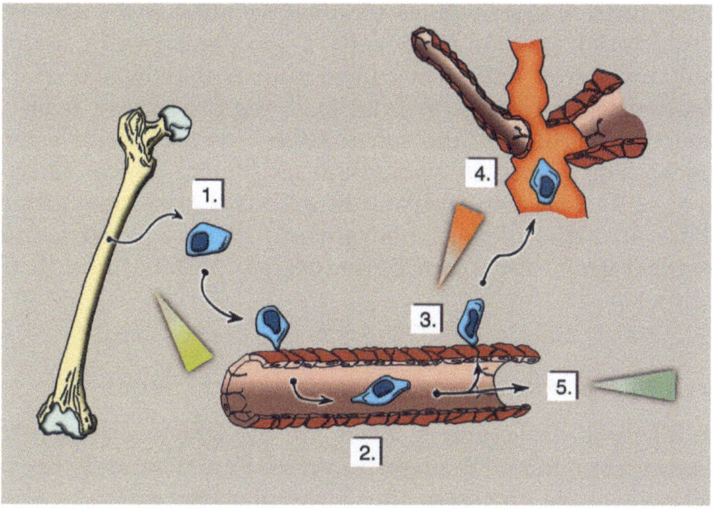

FIGURE 6.1 Several steps in postnatal vasculogenesis. Mobilization from bone marrow to the circulation (1), migration to injury zone (2, 5), invasion in the tissue (3), differentiation in endothelial cells (4), and sprouting (5). Chemokine/signal gradients are represented by triangles. Source: *Courtesy Gaelle Recher.*

into mature ECs or regulate the ECs via paracrine signals (Step 4) [14]. These four steps are required in the damaged endothelium reparation and/or in neovessel formation, and are summarized in Fig. 6.1.

Interestingly, it was demonstrated that during these steps of PNV, EPCs react to angiogenic and vasculogenic factors in order to express several integrin subunits at specific times after entry from the blood circulation into target sites [14]. For example, it was described that integrins $\alpha4$ and $\beta3$ are responsible of anchoring of EPC in BM microenvironment [20]. Later, EPC homing to vascular injury sites is regulated by expression of integrins $\beta2$ and $\beta1$. Integrin $\beta2$ is responsible of EPCs–ECs and EPCs–extracellular matrix (ECM) interactions via VCAM-1 and fibronectin binding, and is also involved in transendothelial migration [21].

Arteriogenesis

Arteriogenesis through vessel enlargement is stimulated under several circumstances. These include increase of arterial pressure, the production of cytokines or growth factors [5]. The vessel has an adaptive response to shear stress which leads to vessel enlargement in order to normalize the biomechanical constraints. The enlargement is due to molecular mechanisms including upregulation of fibroblast growth factors or chemokines [5].

Another form of arteriogenesis is the formation of a collateral circulation by arterial assembly. Recently a mechanisms of arteriogenesis has been uncovered and it has been shown that in neonatal mouse hearts submitted to injury, arterial ECs migrated away from arteries along existing capillaries and reassembled into collateral arteries [6]. The mechanism involves chemokine signaling since arterial ECs express CXCR4 and the capillary ECs CXCL12 which will stimulate arterial EC attraction. It is a mechanism that is required for neonatal heart regeneration, but it can be induced in adults through CXCL12 application.

Postnatal vascularization in regeneration of tissues

From all these observations, EPCs are thought to play a role in the maintenance of endothelial integrity/vessel formation and may regulate PNV in regenerative processes after a vascular injury.

For example, PNV was described as a major contributor to neovascularization after ischemia [15]. Another study demonstrated that an increase in EPC levels after acute ischemic stroke was correlated with a reduction of infarct growth [22]. Moreover, patients with cardiovascular injuries present higher levels of EPCs compared to healthy patients [23] and this was also observed in patients with cerebrovascular diseases [24].

EPC-based therapy may be useful for patients that need neovascularization after cardiovascular diseases. However, the EPC purification and expansion must be improved as well as their delivery to sites of therapeutic interest.

Critical assessment of the importance of postnatal vasculogenesis versus angiogenesis

As mentioned above, there has been an abundant literature on the significant role of EPCs in normal and in pathological vascular development [25–27].

For example, in some tumors, an accumulation of up to 40% of EPCs mobilized from the BM have been reported [25,26]. However, a controversy arose about the validity of these studies. Purhonen et al. performed a systematic study using a variety of mouse models and experimental approaches, but could not observe significant incorporation of EPCs into nascent vessels [7]. This has casted a doubt of the validity of EPC research, at least for the physical participation of EPC in the construction of blood vessels. At minima, EPCs may provide growth and survival factors to stimulate vessel growth.

Furthermore, a second debate has been crystallized around ocular vascular development. Studies reported that choroidal, hyaloid, and retinal vessels are assembled through a vasculogenic process [28–32]. Indeed, spindle-shaped SNA-isolectin positive cells are detected in the mouse retina prior to angiogenic sprout formation, and these cells have been claimed to represent EPCs. However, in 2002 Fruttiger performed a morphological study using antibody labeling and in situ hybridization [31]. Spindle-shape cells did not express VEGFR1/2 but were positive for PDGFRα which represents a marker for immature astrocytes. Thus would indicate that postnatal mouse retinal development exclusively relies on angiogenic sprouting. However, this view has been challenged by studies in mouse, dog, and men [28,33,34]. In the mouse, a novel population of postnatal vascular precursor cells has been identified which expresses the Schwann cell protein myelin protein zero (Po) and are CD45(−)CD31(−)VEcad(−)c-kit(+)CXCR4(+) [33]. These cells are recruited through CXCL12–CXCR4 signaling axis and also require Notch and VEGF signaling. These data indicate that a part of the retinal vascularization (even if this represents minor population) is formed through vasculogenesis [33]. Furthermore, in dog and human retinas, vasculogenesis seems to be the predominant mode of vessel development in the eye [28]. It is to mention that dog and human retinal development proceeds prenatally which is different from the mouse (the development of which occurs during the postnatal period).

Finally, as mentioned above, arteriogenic PNV by vessel assembly is involved in the formation of the neonate collateral circulation [35] and might also occur later in postnatal life and during pathology.

What are the conclusions, we can derive from these studies?

- Postnatal vascular development is not exclusively angiogenesis driven but may also involve vasculogenesis, albeit as a minor component.
- The mouse retina seems to depend primarily on angiogenesis, albeit a minor population of the vasculature is formed through vasculogenesis. In the dog and human retina, vasculogenesis seems to be the leading phenomenon, albeit retinal vascular development proceeds prenatally and not postnatally as in mice.
- Pathological vascular development, such as observed in tumors, seems to depend primarily on angiogenesis, albeit EPC, or bone-marrow–derived cells (BMDCs) can contribute by providing growth and survival factors.
- Arteriogenesis is a specialized form of vasculogenesis and can occur during postnatal life either as a physiological process or as an adaptive response to pathological conditions (i.e., hypertension).

Artificial postnatal vasculogenesis

Endothelial cell–based therapy

Autologous or heterologous graft of EPCs in patients harboring cardiovascular injury is a potential therapeutic strategy but this requires improvements of isolation and culture techniques. Techniques to raise the number of endogenous EPCs, to stabilize the cell population and increase their vasculogenic potential may improve the therapeutic potential of EPC-based therapies. In 2004 Laufs et al. showed that physical activity was associated with a raise of EPC population (CD34 +, VEGFR-2 +) [36]. Mice presented an increase in neovascularization and a decrease of neointima formation after carotid injury. Moreover, an in vivo study demonstrated that EPC treatment was correlated with a better long-term outcome in mice after transient occlusion of the mean cerebral artery and with a reduced infarct volume [37]. Furthermore, another study demonstrated that BM-derived cells participate in the formation of new vessels which is increased by physical activity [38].

It parallel, it was demonstrated that wound healing and myocardial infarction are associated with hypoxia, and an increase in free radicals and inflammation. In order to improve EPC therapy, cytokines such as interleukin-10 (IL-10) significantly increased the vasculogenic potential of EPCs [14]. IL-10 treatment enhanced the survival of EPCs and improved vascularization when compared with untreated EPCs, which confirms vasculogenic potential of this cytokine.

Moreover, it was shown more recently that hypoxia (2%–5%) in isolated EPCs produces reactive oxygen species which upregulates proteases such as matrix metalloproteinase 1. The latter degrades the ECM and promotes formation of EPCs clusters and neovascularization [39]. This is the first description of hypoxia-mediated vasculogenesis, and it will be helpful for improving PNV for tissue regeneration.

Thus EPC therapy associated with exercise, IL-10 or hypoxia-exposure may represent a potential strategy for inducing PNV for improving tissue regeneration after injury.

Therapeutic angiogenesis using transplantation of BMDCs (containing EPCs) was one of the first reports in the setting of a clinical trial (the TACT). In this study, the efficacy of intramuscular injection of autologous BM mononuclear cells in diabetic patients with chronic nonhealing ulcer was reported [40]. Kawamoto et al. also reported the efficacy of autologous and purified CD34 + cells transplant in patients with chronic ischemia in the lower extremities [41], while Matoba et al. demonstrated that cell therapy using BM mononuclear cells can induce a long-term improvement in limb ischemia patients [42].

However, the therapeutic benefit of EPC therapy is subjected to several factors (EPCs purification, amplification, homing at targeted sites, etc.) and may vary. This may limit the use EPCs as a standardized therapy in the clinical setting.

Vascular tissue engineering

Another line of investigation relies on the production of artificial vessel in vitro by vascular tissue engineering. A key step is to produce artificial well-organized multicellular tubular tissue which can be used for vessel grafting. These artificial vessels must exhibit (1) an organized histological two-layered structure comprising a layer of smooth muscle cells (SMCs) enveloping a layer of ECs around a lumen, (2) functional properties such as contractility upon relevant stimuli, and (3) perfusability.

Numerous technologies have been developed such as sheet rolling with ECs, SMCs, and fibroblasts cultured several weeks layer by layer and superimposed as sheets before rolling into luminized tubes [43]. Direct scaffolding is another approach by depositing cellular layers around a tubular mold which is then removed in order to obtain a lumen [44]. A major issue with most of these techniques is the diameter of the produced artificial vessels, because they are more suitable for large vessels (diameter >4 mm) [45].

We recently developed a 3D microfluidic device that allows us to synthesize artificial microvessels (diameter around 400 μm) (Patent WO2018162857 and Ref [46]). We modified the "alginate capsule technology" [47] in order to obtain hollow tubes where SMCs are anchored to an ECM in contact with the alginate shell and where a tight-endothelium in contact with the SMC layer faces the lumen.

The process is summarized as follows. A three-way microfluidic chip is used to inject three solutions simultaneously but at different flow rates inside a calcium bath. The first solution is an alginate solution, the second is an intermediate sorbitol (isotonic) solution, and in the middle of the chip, the third solution is a mix of ECs, SMCs, and matrigel. Calcium is a reticulation agent of alginate, and when injected in a calcium bath, the three solutions gave rise to a tubular shell of a porous alginate hydrogel, with a bulk of cells and matrigel inside. With this new technology, we demonstrated that cells rapidly (24 h) self-organized in a tubular tissue with SMCs anchored to the matrigel in contact with the alginate and a tight endothelium inside. We have extensively characterized these structures we named "vesseloids" and investigated their morphological properties, their contractility, their perfusability, and their response to inflammatory stimuli.

In summary, using this technology, we can easily and rapidly obtain artificial vessels that exhibit functionalities similar to normal blood vessels. This represents a promising tool that may be developed into a therapeutic device for PNV useful for tissue regeneration in many pathological settings.

Conclusion

There is an intricated link between angiogenesis, vasculogenesis, and arteriogenesis. There is controversy of the importance of vasculogenesis in the adult, but it seems evident that this phenomenon can occur during postnatal life. PNV can be activated under specific conditions such as interleukin or chemokine-stimulation or by hypoxia and may represent a therapeutic solution for tissue regeneration. Arteriogenesis is a specific form of PNV which can occur postnatally and during adult life.

The therapeutic benefit of EPC therapy is subjected to several factors which may limit the use as a standardized therapy in the clinical setting.

Research conducted by many laboratories using a tissue engineering approach may contribute to the development of additional therapeutic strategies to improve postnatal vascularization.

Acknowledgements

This work was supported by the ANR VascTubes (ANR-15-CE18-0019) and MecaTiss (ANR-7-CE30-0007-03) to A.B. and P.N. L.A. was supported by the Foundation Lefoulon Delalande, Institut de France. G.R. is a member of the CNRS ImaBio GdR.

References

[1] Patan S. Vasculogenesis and angiogenesis. Cancer Treat Res 2004;117:3–32.
[2] Patan S. Vasculogenesis and angiogenesis as mechanisms of vascular network formation, growth and remodeling. J Neurooncol 2000;50:1–15.
[3] Asahara T, et al. Isolation of putative progenitor endothelial cells for angiogenesis. Science 1997;275:964–7.
[4] Asahara T, et al. Bone marrow origin of endothelial progenitor cells responsible for postnatal vasculogenesis in physiological and pathological neovascularization. Circ Res 1999;85:221–8.
[5] Simons M, Eichmann A. Molecular controls of arterial morphogenesis. Circ Res 2015;116:1712–24.
[6] Das S, et al. A unique collateral artery development program promotes neonatal heart regeneration. Cell 2019;176 1128–1142.e18.
[7] Purhonen S, et al. Bone marrow-derived circulating endothelial precursors do not contribute to vascular endothelium and are not needed for tumor growth. Proc Natl Acad Sci USA 2008;105:6620–5.
[8] Drake CJ. Embryonic and adult vasculogenesis. Birth Defects Res Part C Embryo Today Rev 2003;69:73–82.

[9] Thurston G. Complementary actions of VEGF and angiopoietin-1 on blood vessel growth and leakage. J Anat 2002;200:575–80.
[10] Fong G-H, Rossant J, Gertsenstein M, Breitman ML. Role of the Flt-1 receptor tyrosine kinase in regulating the assembly of vascular endothelium. Nature 1995;376:66–70.
[11] Sato TN, et al. Distinct roles of the receptor tyrosine kinases Tie-1 and Tie-2 in blood vessel formation. Nature 1995;376:70–4.
[12] Adair TH, Montani J-P. Overview of angiogenesis; 2010. Available from: https://www.ncbi.nlm.nih.gov/books/NBK53238/.
[13] Chopra H, Hung MK, Kwong DL, Zhang CF, Pow EHN. Insights into endothelial progenitor cells: origin, classification, potentials, and prospects. Stem Cells Int 2018;2018:1–24.
[14] Balaji S, King A, Crombleholme TM, Keswani SG. The role of endothelial progenitor cells in postnatal vasculogenesis: implications for therapeutic neovascularization and wound healing. Adv Wound Care 2013;2:283–95.
[15] Ribatti D, Vacca A, Nico B, Roncali L, Dammacco F. Postnatal vasculogenesis. Mech Dev 2001;100:157–63.
[16] Khan SS, Solomon MA, McCoy JP. Detection of circulating endothelial cells and endothelial progenitor cells by flow cytometry. Cytom Part B Clin Cytom 2005;64B:1–8.
[17] Miller-Kasprzak E, Jagodziński PP. Endothelial progenitor cells as a new agent contributing to vascular repair. Arch Immunol Ther Exp (Warsz) 2007;55:247–59.
[18] Dong C, Goldschmidt-Clermont PJ. Endothelial progenitor cells: a promising therapeutic alternative for cardiovascular disease. J Interv Cardiol 2007;20:93–9.
[19] Urbich C, et al. Soluble factors released by endothelial progenitor cells promote migration of endothelial cells and cardiac resident progenitor cells. J Mol Cell Cardiol 2005;39:733–42.
[20] Qin G, et al. Functional disruption of α4 integrin mobilizes bone marrow–derived endothelial progenitors and augments ischemic neovascularization. J Exp Med 2006;203:153–63.
[21] Chavakis E, et al. Role of β2-integrins for homing and neovascularization capacity of endothelial progenitor cells. J Exp Med 2005;201:63–72.
[22] Sobrino T, et al. The increase of circulating endothelial progenitor cells after acute ischemic stroke is associated with good outcome. Stroke 2007;38:2759–64.
[23] Yip H-K, et al. Level and value of circulating endothelial progenitor cells in patients after acute ischemic stroke. Stroke 2008;39:69–74.
[24] Liman TG, Endres M. New vessels after stroke: postischemic neovascularization and regeneration. Cerebrovasc Dis 2012;33:492–9.
[25] Kopp H-G, Ramos CA, Rafii S. Contribution of endothelial progenitors and proangiogenic hematopoietic cells to vascularization of tumor and ischemic tissue. Curr Opin Hematol 2006;13:175–81.
[26] Kaplan RN, Rafii S, Lyden D. Preparing the soil: the premetastatic niche. Cancer Res 2006;66:11089–93.
[27] Bertolini F, Shaked Y, Mancuso P, Kerbel RS. The multifaceted circulating endothelial cell in cancer: towards marker and target identification. Nat Rev Cancer 2006;6:835–45.
[28] Lutty GA, McLeod DS. Development of the hyaloid, choroidal and retinal vasculatures in the fetal human eye. Prog Retin Eye Res 2018;62:58–76.
[29] Gariano RF. Cellular mechanisms in retinal vascular development. Prog Retin Eye Res 2003;22:295–306.
[30] McLeod DS, Hasegawa T, Prow T, Merges C, Lutty G. The initial fetal human retinal vasculature develops by vasculogenesis. Dev Dyn 2006;235:3336–47.
[31] Fruttiger M. Development of the mouse retinal vasculature: angiogenesis versus vasculogenesis. Invest Ophthalmol Vis Sci 2002;43:522–7.

[32] Ali Z, et al. Synchronized tissue-scale vasculogenesis and ubiquitous lateral sprouting underlie the unique architecture of the choriocapillaris. Dev Biol 2019. Available from: https://doi.org/10.1016/J.YDBIO.2019.02.002.
[33] Kubota Y, et al. Isolation and function of mouse tissue resident vascular precursors marked by myelin protein zero. J Exp Med 2011;208:949–60.
[34] Ávila-García M, García-Sánchez G, Lira-Romero E, Moreno-Mendoza N. Characterization of progenitor cells during canine retinal development. Stem Cells Int 2012;2012:675805.
[35] Lucitti JL, et al. Formation of the collateral circulation is regulated by vascular endothelial growth factor-A and a disintegrin and metalloprotease family members 10 and 17. Circ Res 2012;111:1539–50.
[36] Laufs U, et al. Physical training increases endothelial progenitor cells, inhibits neointima formation, and enhances angiogenesis. Circulation 2004;109:220–6.
[37] Fan Y, et al. Endothelial progenitor cell transplantation improves long-term stroke outcome in mice. Ann Neurol 2010;67:488–97.
[38] Gertz K, et al. Physical activity improves long-term stroke outcome via endothelial nitric oxide synthase–dependent augmentation of neovascularization and cerebral blood flow. Circ Res 2006;99:1132–40.
[39] Blatchley MR, Hall F, Wang S, Pruitt HC, Gerecht S. Hypoxia and matrix viscoelasticity sequentially regulate endothelial progenitor cluster-based vasculogenesis; 2019. Available from: http://advances.sciencemag.org/.
[40] Sukmawati D, Tanaka R. Introduction to next generation of endothelial progenitor cell therapy: a promise in vascular medicine. Am J Transl Res 2015;7:411–21.
[41] Kawamoto A, et al. Intramuscular transplantation of G-CSF-mobilized $CD34^+$ cells in patients with critical limb ischemia: a phase I/IIa, multicenter, single-blinded, dose-escalation clinical trial. Stem Cells 2009;27:2857–64.
[42] Matoba S, et al. Long-term clinical outcome after intramuscular implantation of bone marrow mononuclear cells (Therapeutic Angiogenesis by Cell Transplantation [TACT] trial) in patients with chronic limb ischemia. Am Heart J 2008;156:1010–18.
[43] L'Heureux N, Pâquet S, Labbé R, Germain L, Auger FA. A completely biological tissue-engineered human blood vessel. FASEB J 1998;12:47–56.
[44] Boccafoschi F, Rajan N, Habermehl J, Mantovani D. Preparation and characterization of a scaffold for vascular tissue engineering by direct-assembling of collagen and cells in a cylindrical geometry. Macromol Biosci 2007;7:719–26.
[45] Song H-HG, Rumma RT, Ozaki CK, Edelman ER, Chen CS. Vascular tissue engineering: progress, challenges, and clinical promise. Cell Stem Cell 2018;22:340–54.
[46] Andrique L, Recher G, Alessandri K, Pujol N, Feyeux M, Bon P, et al. A model of guided cell self-organization for rapid and spontaneous formation of functional vessels. Sci Adv 2019;5(1–12):eaau6562.
[47] Alessandri K, et al. Cellular capsules as a tool for multicellular spheroid production and for investigating the mechanics of tumor progression in vitro. Proc Natl Acad Sci USA 2013;110:14843–8.

CHAPTER

7

The perivascular niche

Alia Komsany and Francesco Pezzella

Nuffield Division of Laboratory Science, Radcliffe Department of Medicine, University of Oxford, John Radcliffe Hospital, Oxford, United Kingdom

Introduction

The "seed and soil" hypothesis introduced by Steven Paget was a pivotal discovery in the study of malignant disease, and in particular, our understanding of metastasis [1]. This hypothesis put forward the notion of a nutritive microenvironment or niche as a prerequisite for the dissemination and subsequent metastases of tumor cells to distant tissue sites. Preceding this, the theory that metastatic tumor dissemination was solely determined by the embedding of tumor cell emboli within the vascular network, prevailed. From Paget's analysis of 735 cases of breast cancer, however, he concluded that a select number of organs such as the liver, appeared to be specifically prone to metastases, and that this particular tendency could not be attributed to blood flow alone [2]. He thus deduced that it was the soil or local tissue microenvironment of these organs that induced the seeding of disseminated tumor cells and promoted metastases at select distant organ sites in favor of others (such as the spleen). Forty years on, Paget's hypothesis came under scrutiny and was challenged by James Ewing, who once again propositioned that it was the vascular framework and lymphatic drainage network that determined the metastatic potential of tumor cells [2]. This view prevailed until formative studies were carried out by Isaiah Josh Fidler, categorically highlighting that although tumor cells reach the vasculature of all organs, it was only in select organs that metastases was shown to occur, while in others, it did not [3].

Tumor Vascularization

DOI: https://doi.org/10.1016/B978-0-12-819494-2.00007-9

A renewed focus on the metastatic soil hypothesis thus ensued along with extensive research, investigating the pathophysiology of the local tissue microenvironment or niche of tumor cells from both the primary tumor and metastatic sites [4].

What is a perivascular niche?

The term "niche" is used in biology to describe an environment and its interaction with a given population [5]. Originally a concept used for ecological environmental studies, the term is now used to describe the microenvironment inside a body made up of anatomical structures and a preexisting population of cells. The perivascular niche describes the microenvironment(s) around a vessel and includes distinct cell types and stroma, involved in the release of specific cues, critical for survival. In physiological conditions, the perivascular niche is the microenvironment which harbors postnatal stem cells [6]. As a result, the hypothesis, that these microenvironments are also essential for initial survival, invasion, proliferation, stemness, and the subsequent resistance of cancer stem cells, has been put forward. One of the best studied types of niche is the one hosting the hemopoietic stem cells (HSCs) in the murine bone marrow and its main features are illustrated in Fig. 7.1.

In this specialized anatomical, physiological, and molecular space, tumor cells can find optimal access to oxygen and nutrients from the blood stream. However, this understanding is far from the full picture, as it has also been demonstrated that neoplastic metastatic cells need to reside in this perivascular niche in order to survive and thrive, despite the fact that oxygen and nutrient availability can be provided from neighboring blood vessels.

What characterizes these microanatomical structures is not only the proximity to a vessel but also the presence of a variety of cell populations such as pericytes, endothelial cells, immune cell subtypes, and others, which tend to be organ specific (e.g., astrocytes in the brain).

Furthermore, cells in the extracellular matrix including the basement membrane of the blood vessel as well as the immunological response need to be specifically tailored [4].

Cancer cells that enter an organ and fail to localize, die. Whereas those that settle in these niches will either keep growing or become dormant. This difference in behavior is not only determined by the phenotype/genotype of the neoplastic cell but also may be dependent on the type of niche; the cancer cells have selected to settle in. It is therefore very likely that the specific conditions of a perivascular space could be the deciding factor in a tumor cell's destiny.

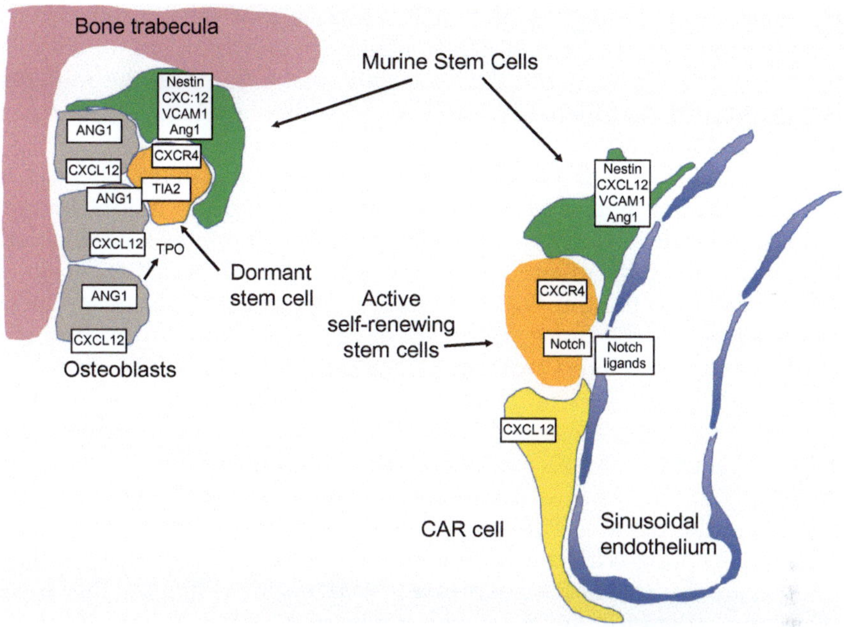

FIGURE 7.1 **An example of perivascular niche: HSC in murine bone marrow.** The niche for the murine HSC is probably the one we know more about. Situated in the bone marrow, there is a paratrabecular and a perivascular component. The paratrabecular niche main components are the osteoblasts and the mesenchymal stem cells nestin positive among which HSCs are located. The perivascular section is composed by the same mesenchymal stem cells nestin positive, by the CXCL12-abundant reticular (CAR) cells, and the very same endothelial cells. The CAR cells are so far regarded as possible progenitors of the osteogenic cells. The CAR cells could possibly derive from the mesenchymal stem cells nestin positive. The hemopoietic cells in the paratrabecular niche are maintained quiescent by different mediators like thrombopoietin, angiopoietin 1 (also involved in maintaining quiescent the endothelium), CXCL12, and adhesion molecules like VCAM. In the perivascular region instead, the presence of CAR cells and of endothelial cells triggers the hemopoietic cells into self-renewal. The endothelial cells induce it by presenting NOTCH ligands to the stem cell, and the CAR cells are known to be essential for the hemopoietic cells to cycle although the mechanism is not yet fully understood. Source: *Based on* [6–9].

Preparing the soil: the premetastatic niche

In addition to the aforementioned studies, other pivotal discoveries have shown that tumors can initiate the development of microenvironments located in distant organs that help contribute to the survival and dissemination of tumor cells prior to their arrival at these sites. These studies have led to the creation of another concept, the "premetastatic niche." As cancer cells differ in variable degrees from normal ones, not necessarily the physiological niche would allow the cancer cell survival [4]. In line with both Paget's and Ewing's theories, the concept of the premetastatic niche differs in that it suggests that the primary tumor initiates a preconditioning of organ sites prior to the arrival of circulating tumor cells. In contrast, the metastatic niche involves the initiation and shaping of the microenvironment once the circulating tumor cells have arrived [10,11].

The original paper from Kaplan et al. [11] provides an excellent illustration of the premetastatic niche concept. In a murine model injected with bone marrow–derived cells (BMDCs) also known as hematopoietic progenitor cells (HPCs) and either Lewis lung carcinoma (LLC) cells or B16 melanoma cells, the BMDCs were found to colonize the premetastatic sites before the cancer cells. These sites were found to be different for LLC (lung and liver) from those for melanoma (lung, liver, spleen, testis, and kidney). The BMDC found in the premetastatic sites expressed VEGFR1, these cells are also known as HPCs. Depletion of VEGFR1 positive BMDC or treatment with anti-VEGFR1 antibodies abrogates the formation of the metastatic lesions. The BMDCs are able to colonize the niche, expressing VLA-4 which mediates their adhesion to the niche, along with the expression of alpha4beta7 and alpha6beta4.

Subsequently the BMDC expresses MMP9 which initiates the breakdown and alteration of the local basal membrane, resulting in the release of VEGFA and cKit which facilitate new incoming cells to survive. In this modified niche containing VEGFR1 positive cells, fibroblasts and fibronectin, there is an overexpression of CXCL12 which provides a gradient attracting CXCR4 positive tumor cells [11].

Since than an increasing body of work is being investigating the formation of premetastatic niches. This line of investigation has utilized mostly orthotopic and transgenic mouse models of lung metastasis, with lung metastasis being a strong focus, owing to the fact that it is the most common site of metastasis in both preclinical and patient models [4,12].

Organotropism: organ-specific premetastatic niches

Following the original observation of Kaplan [13], an increasing body of clinical evidence points to the existence of premetastatic niches present in cancer patient tissue samples and that different tumors utilize different types of niche. Premetastatic niches have been found in the lymph nodes of patients with colorectal, prostrate, breast, thyroid, bladder, gastric, and renal cell carcinomas [14].

Lymph node premetastatic niches were observed both in mouse models and in patients with breast cancer, highlighting the expression of provasculogenic vascular endothelial growth factor 1 myeloid progenitor cells that inhabit lymph nodes primed for the engraftment of circulating tumor cells by tumor secreted factors [15]. Furthermore, it was demonstrated that lymph angiogenesis occurs prior to circulating tumor cell arrival at distant sites of lymph node metastasis [16]. Upon encountering immune cells within the lymph nodes, tumor cells can interact either directly or indirectly with these immune cells by means of tumor secreted factors or extravascular vesicles [17]. These interactions are thought to regulate immune responses that act to oppose circulating tumor cells [18].

Premetastatic niche formation in the liver (occurs at the initial phases of visceral metastasis) has also been observed to rely on BMDC population enlistment [19]. Enzymes involved in extracellular matrix remodeling have been found to produce liver premetastatic niches by means of stromal cell-derived factor 1 dependant recruitment of neutrophils [20]. Furthermore exosomes, specifically those derived from human pancreatic cancer cells expressing macrophage migration inhibitory factor, were also shown to initiate liver premetastatic niches in addition to their role as biomarkers [21]. These exosomes were found to interact with Kupffer cells to produce transforming growth factor B, promoting extracellular matrix remodeling, which in turn facilitated the recruitment of bone marrow–derived macrophages [1].

Similarly, bone is considered to be a common site of metastasis for a variety of solid tumor types [22]. Both primary tumors and the circulating factors produced by them act to precondition select cells present in the bone, to support premetastatic niches and the establishment of metastatic cell foundations [23].

Another premetastatic tissue site is the brain, an organ protected by the blood–brain barrier and one that is considered to be one of the most detrimental metastatic sites associated with treatment failure [24]. Although understanding brain metastasis remains a challenge, alterations in astrocytic basement membrane component laminin-$\alpha 2$ and pericyte subpopulations have been implicated to play a role in the permeability of the blood–brain barrier [25]. In addition to this, glucose metabolism has been suggested as one of the most relevant mechanisms involved in breast cancer metastasis to the brain [26].

Early studies, involving the injection of a lung carcinoma cell line (restricted metastases to the lung) into the flank of a syngeneic mice pretreated with melanoma condition medium have shown that tumor-derived factors in addition to observations made later involving tumor-secreted exosomes, are a vital component in the formation of premetastatic niches [11]. The lung carcinoma cells from mice treated with the melanoma condition medium were found to metastasize to the spleen, kidney, oviduct, and intestine (all organ sites associated with melanoma metastasis [11]). This reinforces the notion that tumor cell intrinsic properties are not sufficient for metastatic seeding and outgrowth and that a preconditioned microenvironment is essential for organ-specific metastasis [1].

Cancer stem cell niche

What is becoming increasingly apparent is that tumors are made up of heterogeneous cellular populations, organized in a hierarchical manner, with cancer stem cells forming the top fraction of this cellular hierarchy. This top cellular compartment shares many similarities with that of normal stem cells including self-renewal and differentiation [27]. Cancer stem cells are highly tumorigenic and are capable of transplanting themselves in a serially transplantable manner. One of the main issues with cancer stem cells is their resistance to treatment, which is thought to be the underlying cause of cancer recurrence [28].

Like normal stem cells, cancer stem cells reside in distinct environments or niches. These niches are typically made up of diverse stromal cells, which include mesenchymal and immune cells, a vascular network, soluble factors and components of the extracellular matrix. To remain tumorigenic, a complex interplay between tumor cells and nonmalignant cells needs to be maintained. This delicate balance is regulated by the cancer stem cell niche, which enables tumor cells to retain their ability to self-renew and produce progenitor cells while remaining undifferentiated themselves [29].

In addition to this, the niche has a protective role, in that it shields cancer cells from genotoxic insults and is involved in advanced therapeutic resistance [30].

Understanding the conserved biology of stem cells and their respective niches provides insights into the behavior of these niches during homeostasis and tissue repair, and likewise can aid in our understanding of signaling mechanisms in a disease setting, particularly in the context of cancer metastasis and survival.

Perivascular niche: the arrival of the cancer cells

The neoplastic cells found to be nesting in the perivascular niche of an organ can be both from the organ's primary tumor or metastatic cells from another site (Fig. 7.2).

How cancer cells disseminate from the primary tumor to a perivascular niche is still poorly understood. One model using glioma cells has been described by Montana and Sontheimer [31]. They demonstrated that glioma cells express the bradykinin 2 receptor (B2R). As the endothelial cells express bradykinin, a gradient of this molecule is established around the vessels. When the glioma cells meet bradykinin, its receptor is activated inducing intracellular Ca^{2+} oscillations leading to increased kinetic activity and the migration of the neoplastic cells. Cells

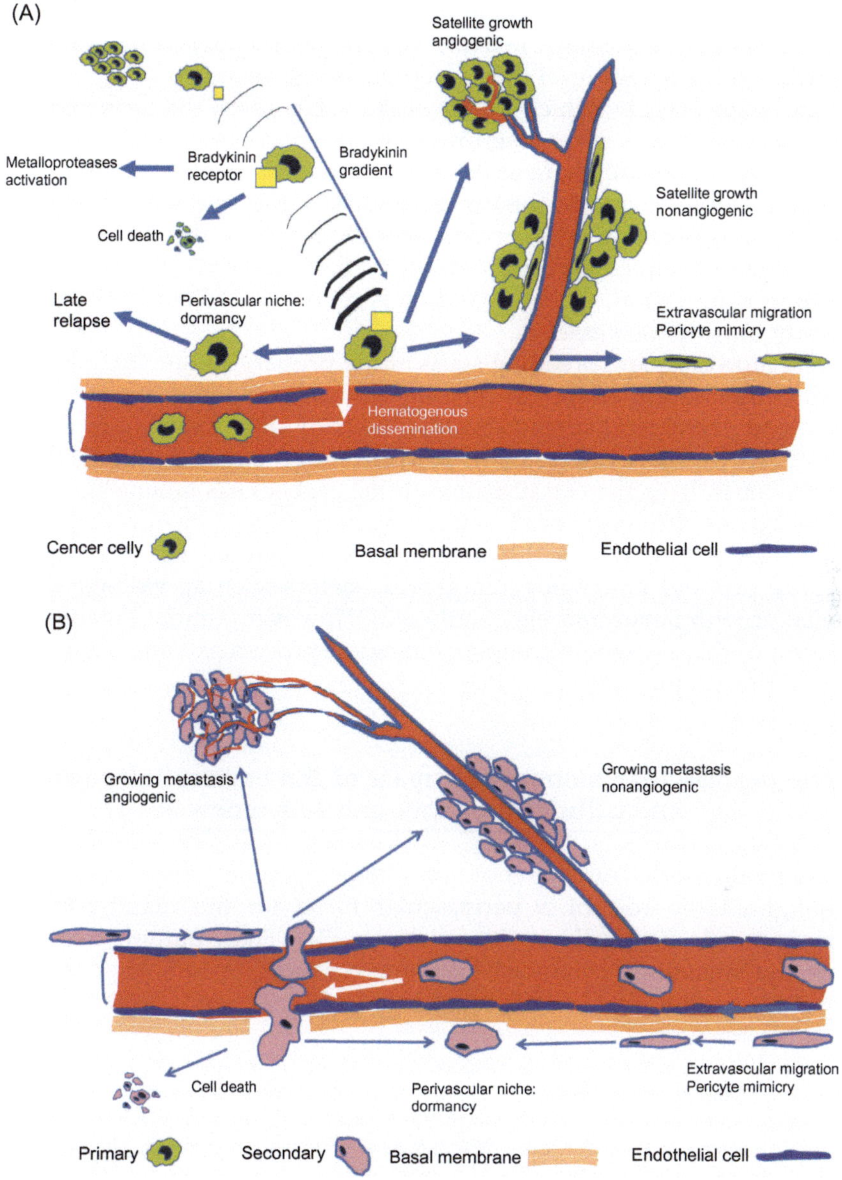

FIGURE 7.2 Cancer cells destiny after leaving the primary tumor or after extravasation. (A) *Cells spreading from the primary tumor. The glioblastoma model.* As the glioma cells express B2R, they move along the gradient of bradykinin present in the tissue. This gradient originates from the endothelial cells. When the glioma cells meet bradykinin, its receptor is activated inducing intracellular Ca^{2+} oscillations leading to increased kinetic activity and the migration of the neoplastic cells. Furthermore, bradykinin 2 receptor induces production of matrix metalloproteases which further favor the cancer cell

which moving leave the gradient halt to a stop, but cells moving toward the vessels are increasingly activated by the denser concentration of bradykinin and progress until they reach the blood vessel.

The main way, by which the metastatic cells reach the target organs, is by hematogenous spreading through the blood vessels and, less commonly, by lymphatic channels. A third, less common mechanism is by pericyte mimicry, that is, the cancer cells flow the vessel on the abluminal side by co-opting it. When reaching the target by this latter method, described in Chapter 4, Pericyte mimicry: an embryonic-like mechanism for tumor metastasis, no extravasation is required as the cancer cells are already outside the vessels.

The intravascular spreading starts at the primary tumor site with vascular intravasation of the cancer cells to escape the primary site, enter the blood vessel, and spread through the body. It is followed in the target organ by the extravasation. A wide range of factors and cellular components drive the extravasation of the cancer cells among these platelets act as delivery vehicles for a variety of angiogenic regulatory molecules whose role is to alter local vasculature. These molecules include pro and antiangiogenic growth factors such as vascular endothelial growth factor and endostatin [32]. However, a more detailed discussion of this extremely complex biological process extends beyond the aim of this chapter [33].

The perivascular niche: the impact of the extracellular matrix and cellular crosstalk and self-renewal

Seminal work carried out on hematopoietic stem cells have highlighted the role of a perivascular niche for the maintenance of hematopoietic stem cells in vitro and vivo [34]. These authors described how the signatures produced by the expression, or lack of it of three proteins of the SLAM family (CD150, CD244, and C48) identify,

spreading [31]. On reaching the perivascular location, different events can happen according to the phenotype and genotype of the cancer cells and the microenvironment. The cells can remain dormant, if the appropriate niche is present, to produce perhaps later on a relapse. Or they can spread in the classic hematogenous way, by entering the vessels, or by pericyte mimicry crawling on the abluminal surface. They can also produce immediately new neoplastic lesion both by nonangiogenic, angiogenic, or mixed patterns. Cells meeting forbidding conditions will die, usually by apoptosis. (B) *Metastatic cells from other organs*. When a metastatic cancer cell arrives form a distant primary, again different events can follow. After extravasation, if arriving as in most of the cases by hematogenous rather than abluminal spreading, the neoplastic cells can remain dormant if an appropriate niche is present, or can progress immediately into a new metastatic lesion, angiogenic, nonangiogenic, or mixed. Some cancer cells, unable to find the right niche or to grow into a metastasis, will die.

respectively, HSCs are CD150 + /CD244 − /C48 − , the multipotent progenitors are CD150 − /CD244 + /C48 − , and finally, the most restricted progenitors are CD150 − /CD244 + /C48 + . Exploiting these signatures, they described that HSCs are found in the murine spleen in proximity of the sinusoidal endothelium while in the marrow were present in proximity of the marrow sinusoids and of the osteoblasts.

A successive study [35] has demonstrated that these hematopoietic stem cells localize by small arteriole in endosteal bone marrow. These niche contains quiescent mesenchymal cells characterized according to their variable level of nestin expression [35] and which are positive also for CXCL1 and leptin receptor [36]. Also, the HSC nested among these mesenchymal cells have been found to be quiescent. These arterioles are surrounded by a special type of pericyte called NG2, also known as CSPG4, which also express nestin. Depletion of these pericytes leads to a switch of the HSCs from a quiescent to a proliferative status [35]. These studies led to the conclusion that these perivascular niches are composed of a variety of cell types, each with a unique function in the maintenance of hematopoietic stem cells [6,35].

Furthermore, studies conducted on mesenchymal stem cells, nonhematopoietic cells capable of self-renewal along with the ability to give rise to multiple lineages, have highlighted that postnatal mesenchymal stem cell also resides in perivascular niches, have the capacity to differentiate into other cell populations such as pericytes, and are located and maintained near the host vasculature [37].

Stepping into the perivascular niche

Once a cancer cell leaves the primary tumor can reach a niche in the same organ and become either dormant or generate immediately a new lesion. Other way can enter a vessel and be disseminated until extravasate and get itself established in a new organ. However, can also, at any stage, die if the conditions are not adequate (Fig. 7.3A).

A model of dissemination inside the same organ explored by various authors is the glioma. Montana and Sontheimer [31] described in an ex vivo model which is based on maintaining slices of rat brain cortex in a viable state in vitro, how a human glioma cell line expressing the B2R, an activator of matrix metalloproteinases, allows the cells to migrate toward the blood vessels. Caspani et al. [39] unraveled the mechanism which then regulates the interaction of the glioma cells with the vessel. Seeding of glioblastoma cells on mouse brain explants resulted in vascular co-option within 15 hours. The tumor cells developed actin-based cytoplasmic extensions, called "flectopodia," which contact target pericytes. For the process to happened, it is crucial the

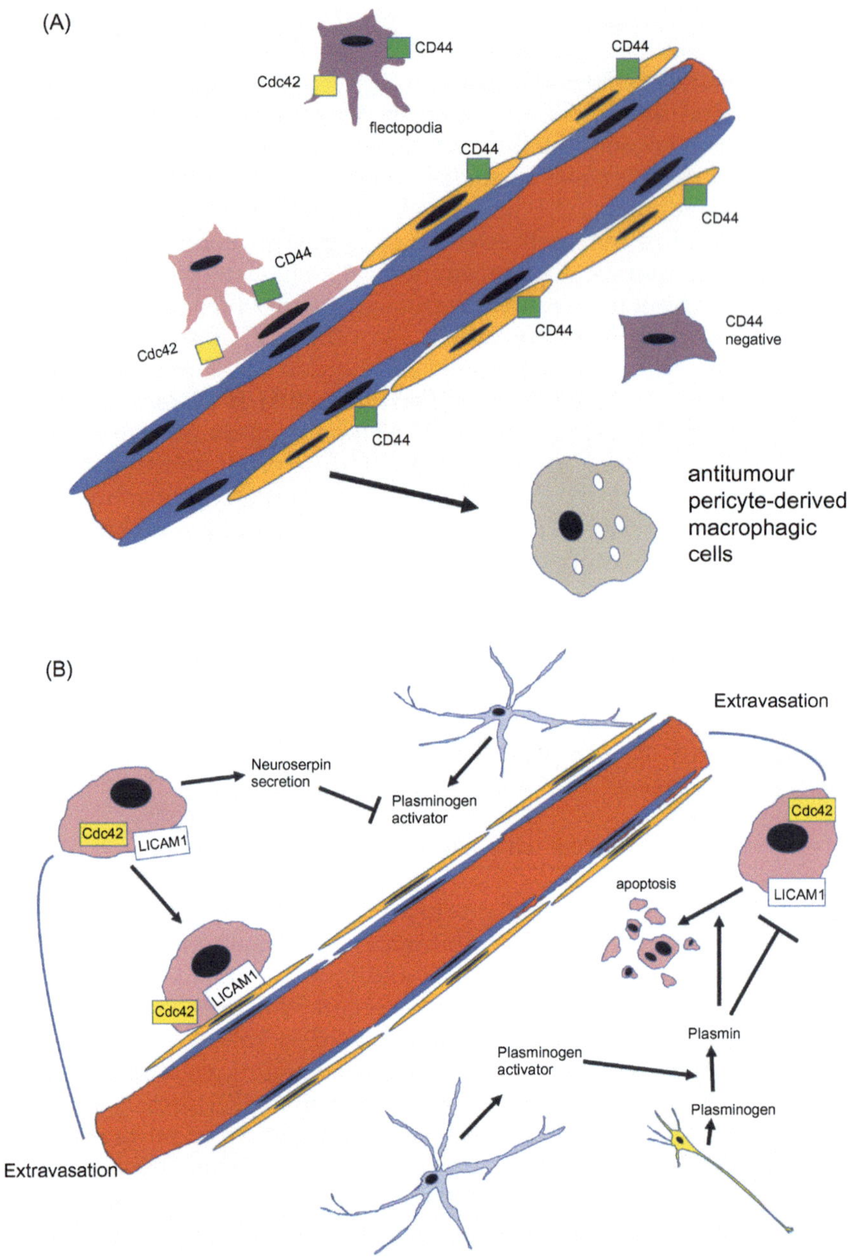

FIGURE 7.3 Survival versus apoptosis. (A) *The glioblastoma model*. Glioblastoma cells expressing Cdc42, an RHO GTPase regulating actin-dependent cytoplasm extensions, and the fusogenic protein CD44 co-opt the vessel by linking to and fusing with the pericytes. If just one of these two proteins are not present, the neoplastic cell fails to co-opt the

expression of CDC42, an RHO GTPase regulating actin-dependent cytoplasm extensions, and CD44, a fusogenic protein.

Silencing of the expression of CDC42 and/or CD44 results in failure of the neoplastic cells to co-opt the vessels and therefore spread the tumor. Instead non-co-opting cancer cells induces metamorphosis of some pericytes into antitumor macrophages which lead to a change of microenvironment unfavorable for the glioma [39].

Finally, Watkins et al. [42] described that glioblastoma cells can also insert under the astrocyte end feet and interrupt coupling of astrocytes to blood vessels and can over the ability of astrocytes to influence vascular tone.

Metastatic cells also have specific perivascular niche in the brain (Fig. 7.3B). If the cell fails to co-opt the vessels after extravasation, they do not proliferate as illustrated by Kienast et al. [43] using multiphoton laser scanning microscopy and CDC42 is likely to be involved also to this process. Reymond et al. [44] showed that CDC42 is essential in the metastatic process not only to allow extravasation but also to co-opt and move along the abluminal surface of the vessels. They show that during extravasation CDC42 expression increases the levels of serum response factor, which then leads to transcription and expression of β1 integrin which allow the metastatic cells to adhere to the endothelium, the first step for extravasation. After intercalation, it is again necessary for the cancer cell to protrude and adhere to the basement membrane to complete extravasation and remain attached to the abluminal surface of the vessel. Direct blockade of β1 integrin subunit prevented adhesion of tumor cells to the subendothelial basement membrane, thereby impairing colonization and therefore the possibility of metastases, the same results were obtained by blocking CDC42 [44].

In another model of murine brain metastases, Valiente et al. [40] dissected the mechanism which facilitates a cancer cell to co-opt the abluminal surface of a vessel. Neurons secret plasminogen, which is cleaved to plasmin by the astrocyte-secreted tissue plasminogen activator (PA) and urokinase PA. Plasmin inactivates the adhesion molecule L1CAM

◀ vessel. When large areas of non-co-opted pericytes are present, some of these cells detach form the basal membrane and convert into macrophagic cells able to eliminate non-co-opted cancer cells [38,39]. (B) *The metastatic carcinoma model*. In this model, cell lines form lung and breast adenocarcinomas were used. On the right, a metastatic cell has entered the brain parenchyma. As astrocyte secrete plasminogen activator, the plasminogen, produced by the neurons, is cleaved into its active form plasmin. Plasmin has two effect on the metastatic cell: blocks the adhesion molecule LICAM1, necessary for vascular co-option, and also activate FASL-mediated apoptosis causing the death of the neoplastic cells. However, see on the left side, if the metastatic cell secretes neuroserpin, plasminogen activator is neutralized. Therefore apoptosis is not triggered, LICAM1 remains active and the cancer cell co-opt the vascular structures and remain viable [40,41].

expressed on the surface of the cancer cells and cleaves the membrane-associated FAS ligand (FASL) expressed on the astrocytes cell membrane. In this way, the soluble form sFASL is released causing the apoptosis of cancer cells. However, they found in their model that, if the extravasated cancer cells have the ability to secrete either neuroserpin or serpin B2, PA is inactivated. Therefore the plasmin-mediated pathway leading to apoptosis of the metastatic cell is interrupted and, if the metastatic cells evading apoptosis are L1CAM positive, they adhere to the basal membrane, co-opt the vessel, and can then eventually produce a metastatic lesion [40]. For such a metastatic lesion to happen by nonangiogenic growth, exploiting preexisting vessels, CDC42 is gain essential as CDC42 negative cells fail to move along the co-opted vessels and proliferate [44].

Perivascular niche: dormancy

It has been observed that distant metastases emerge after a period of latency that can last years if not decades. Unraveling the mystery of how disseminated tumor cells are kept dormant and what exactly wakes them up from this dormancy remains a fundamental problem in tumor biology. Experiments in mouse models of breast cancer have attempted to address and answer these questions. These studies hypothesized that the microvascular basement membrane is first encountered by disseminating tumor cells, the endothelial cells of which play a crucial role in releasing factors that sustain quiescence [45]. It was thus found that dormant tumor cells reside on microvasculature of lung, bone marrow, and brain and was proved that endothelial cells in fact maintain breast cancer cell quiescence.

Proteomic and functional analysis of the proteins that populate organotypic microvascular niches, identified thrombospondin-1 (TSP-1) as an endothelium-derived tumor suppressor. It was also found that expression of TSP-1 was downregulated close to sprouting neovasculature, implying that tumors may be able to escape growth regulation in this subniche. This was further confirmed as growth was shown to be accelerated around neovascular tips which were observed to express active tumor promoting factors [45].

The idea that endothelial cells directly regulate cells in perivascular niche has long been established in a number of biological studies on normal tissues, namely endothelial cells that phenotypically resemble neovascular tip cells induce growth, morphogenesis of the liver [46], and regeneration of the lung alveoli [46]. While established endothelium on the other hand initiates pancreatic differentiation, the inhibition of smooth muscle proliferation, promoting pluripotency of neural, hematopoietic, and mesenchymal stem cells.

The studies carried out on breast cancer dormancy and its relation to perivascular niche, confirm the notion that endothelial subniches can be found to distinctly present differential regulation of disseminated tumor cells. Thus suggesting that the preservation of vascular homeostasis is essential for the sustaining the dormancy of disseminated tumor cells. Furthermore, these studies also demonstrate that the conference of quiescence by mature microvasculature and tissue growth by sprouting endothelium also applies in the context of the disseminated tumor cell microenvironment [45].

Conclusion

Despite the advancements made surrounding the biology and therapeutic response of cancer stem cell niche microenvironments and in particular the perivascular niche, there still remains a lot more to be investigated along with the establishment of improved human clinical models to better predict patient responses. Being able to elucidate the cellular mechanisms and the tumor environment that regulates tumor latency, is crucial in order to develop through targeted therapy which could destroy dormant cells or, at least, prevent them from "waking up."

References

[1] Peinado H, Zhang H, Matei IR, et al. Pre-metastatic niches: organ-specific homes for metastases. Nat Rev Cancer 2017;17:302–17.
[2] Ewing J. Neoplastic diseases. A treatise on tumours. Am J Med Sci 1928.
[3] Fidler IJ, Kripke ML. Metastasis results from preexisting variant cells within a malignant tumor. Science 1977;197:893–5.
[4] Psaila B, Lyden D. The metastatic niche: adapting the foreign soil. Nat Rev Cancer 2009;9:285–93.
[5] Thain M, Hickman M. The penguin dictionary of biology. London: Penguin Books; 2004.
[6] Oh M, Nor JE. The perivascular niche and self-renewal of stem cells. Front Physiol 2015;6:367.
[7] Sugiyama T, Kohara H, Noda M, Nagasawa T. Maintenance of the hematopoietic stem cell pool by CXCL12-CXCR4 chemokine signaling in bone marrow stromal cell niches. Immunity 2006;25:977–88.
[8] Ehninger A, Trumpp A. The bone marrow stem cell niche grows up: mesenchymal stem cells and macrophages move in. J Exp Med 2011;208:421–8.
[9] Zhang H, Bosch-Marce M, Shimoda LA, et al. Mitochondrial autophagy is an HIF-1-dependent adaptive metabolic response to hypoxia. J Biol Chem 2008;283:10892–903.
[10] Kaplan RN, Psaila B, Lyden D. Bone marrow cells in the 'pre-metastatic niche': within bone and beyond. Cancer Metastasis Rev 2006;25:521–9.

[11] Kaplan RN, Riba RD, Zacharoulis S, et al. VEGFR1-positive haematopoietic bone marrow progenitors initiate the pre-metastatic niche. Nature 2005;438:820–7.
[12] Aguado BA, Bushnell GG, Rao SS, Jeruss JS, Shea LD. Engineering the pre-metastatic niche. Nat Biomed Eng 2017;1.
[13] Kimura E, Enns RE, Alcaraz JE, Arboleda J, Slamon DJ, Howell SB. Correlation of the survival of ovarian cancer patients with mRNA expression of the 60-kD heat-shock protein HSP-60. J Clin Oncol 1993;11:891–8.
[14] Sleeman JP. The lymph node pre-metastatic niche. J Mol Med (Berl) 2015;93:1173–84.
[15] Karaca Z, Tanriverdi F, Unluhizarci K, et al. VEGFR1 expression is related to lymph node metastasis and serum VEGF may be a marker of progression in the follow-up of patients with differentiated thyroid carcinoma. Eur J Endocrinol 2011;162:277–84.
[16] Hirakawa S, Brown LF, Kodama S, Paavonen K, Alitalo K, Detmar M. VEGF-C-induced lymphangiogenesis in sentinel lymph nodes promotes tumor metastasis to distant sites. Blood 2007;109:1010–17.
[17] Jung T, Castellana D, Klingbeil P, et al. CD44v6 dependence of premetastatic niche preparation by exosomes. Neoplasia 2009;11:1093–105.
[18] Headley M, Bins A, Nip A, et al. Visualization of immediate immune responses to pioneer metastatic cells in the lung. Nature 2016;531:513–17.
[19] Zhang C, Zhou C, Wu XJ, et al. Human CD133-positive hematopoietic progenitor cells initiate growth and metastasis of colorectal cancer cells. Carcinogenesis 2014;35:2771–7.
[20] Seubert B, Grünwald B, Kobuch J, et al. Tissue inhibitor of metalloproteinases (TIMP)-1 creates a premetastatic niche in the liver through SDF-1/CXCR4-dependent neutrophil recruitment in mice. Hepatology 2015;61:238–48.
[21] Costa-Silva B, Aiello NM, Ocean AJ, et al. Pancreatic cancer exosomes initiate pre-metastatic niche formation in the liver. Nat Cell Biol 2015;17:816–26.
[22] Mundy GR. Metastasis to bone: causes, consequences and therapeutic opportunities. Nat Rev Cancer 2002;2:584–93.
[23] Cox TR, Bird D, Baker AM, et al. LOX-mediated collagen crosslinking is responsible for fibrosis-enhanced metastasis. Cancer Res 2013;73:1721–32.
[24] Steeg PS, Camphausen K, Smith QR. Brain metastases as preventive and therapeutic targets. Nat Rev Cancer 2011;11:352–63.
[25] Lyle LT, Lockman PR, Adkins CE, et al. Alterations in pericyte subpopulations are associated with elevated blood-tumor barrier permeability in experimental brain metastasis of breast cancer. Clin Cancer Res 2016;22:5287–99.
[26] Fong MY, Zhou W, Liu L, et al. Breast-cancer-secreted miR-122 reprograms glucose metabolism in premetastatic niche to promote metastasis. Nat Cell Biol 2015;17:183–94.
[27] Clarke MF, Dick JE, Dirks PB, et al. Cancer stem cells—perspectives on current status and future directions: AACR Workshop on cancer stem cells. Cancer Res 2006;66:9339–44.
[28] Todaro M, Alea MP, Di Stefano AB, et al. Colon cancer stem cells dictate tumor growth and resist cell death by production of interleukin-4. Cell Stem Cell 2007;1:389–402.
[29] Calabrese C, Poppleton H, Kocak M, et al. A perivascular niche for brain tumor stem cells. Cancer Cell 2007;11:69–82.
[30] Hovinga KE, Shimizu F, Wang R, et al. Inhibition of notch signaling in glioblastoma targets cancer stem cells via an endothelial cell intermediate. Stem Cells 2010;28:1019–29.
[31] Montana V, Sontheimer H. Bradykinin promotes the chemotactic invasion of primary brain tumors. J Neurosci 2011;31:4858–67.

[32] Schlesinger M. Role of platelets and platelet receptors in cancer metastasis. J Hematol Oncol 2018;11:125.
[33] Chitty JL, Filipe EC, Lucas MC, Herrmann D, Cox TR, Timpson P. Recent advances in understanding the complexities of metastasis. F1000Res 2018.
[34] Kiel MJ, Yilmaz O, Iwashita T, Yilmaz OH, Terhorst C, Morrison SJ. SLAM family receptors distinguish hematopoietic stem and progenitor cells and reveal endothelial niches for stem cells. Cell 2005;121:1109–21.
[35] Kunisaki Y, Bruns I, Scheiermann C, et al. Arteriolar niches maintain haematopoietic stem cell quiescence. Nature 2013;502:637–43.
[36] Greenbaum A, Hsu YM, Day RB, et al. CXCL12 in early mesenchymal progenitors is required for haematopoietic stem-cell maintenance. Nature 2013;495:227–30.
[37] Zhao H, Feng J, Seidel K, et al. Secretion of shh by a neurovascular bundle niche supports mesenchymal stem cell homeostasis in the adult mouse incisor. Cell Stem Cell 2014;14:160–73.
[38] Caspani EM, Echevarria D, Rottner K, Small JV. Live imaging of glioblastoma cells in brain tissue shows requirement of actin bundles for migration. Neuron Glia Biol 2006;2:105–14.
[39] Caspani EM, Crossley PH, Redondo-Garcia C, Martinez S. Glioblastoma: a pathogenic crosstalk between tumor cells and pericytes. PLoS One 2014;9:e101402.
[40] Valiente M, Obenauf AC, Jin X, et al. Serpins promote cancer cell survival and vascular co-option in brain metastasis. Cell 2014;156:1002–16.
[41] Valiente M, Ahluwalla MS, Boire A, et al. The evolving landscape of brain metastasis. Trends Cancer 2018;4:176–96.
[42] Watkins S, Robel S, Kimbrough IF, Robert SM, Ellis-Davies G, Sontheimer H. Disruption of astrocyte–vascular coupling and the blood–brain barrier by invading glioma cells. Nat Commun 2014;5.
[43] Kienast Y, von Baumgarten L, Fuhrmann M, et al. Real-time imaging reveals the single steps of brain metastasis formation. Nat Med 2010;16:116–22.
[44] Reymond N, Im JH, Garg R, et al. Cdc42 promotes transendothelial migration of cancer cells through β1 integrin. J Cell Biol 2012;199:653–68.
[45] Ghajar CM, Peinado H, Mori H, et al. The perivascular niche regulates breast tumour dormancy. Nat Cell Biol 2013;15:807–17.
[46] Rafii S, Butler JM, Ding B-S. Angiocrine functions of organ-specific endothelial cells. Nature 2016;529:316–25.

CHAPTER

8

Zebrafish embryo as an experimental model to study tumor angiogenesis

Jessica Guerra, Chiara Tobia, Marco Presta and Andrea Barbieri
Department of Molecular and Translational Medicine, University of Brescia, Brescia, Italy

Zebrafish as a model for cancer research

Zebrafish (*Danio rerio*) is a small fresh-water fish that emerged in the past decades as a model for the study of various diseases. Easy manipulation and low costs of maintenance contributed to its diffusion through research laboratories over the world [1]. Due to the evolutionary conservation between zebrafish and human genomes [2], molecular pathways leading to the onset and progression of human cancer can be studied in zebrafish, making this animal platform a useful tool in tumor research [3].

Several approaches have been used to induce cancer in zebrafish. Initially, zebrafishes were treated with common mutagens (e.g., ethylnitrosourea or 7,12-dimethylbenz[*a*]anthracene) at different stages of development, leading to the formation of cancer lesions in various organs including skin, liver, gill, blood vessels, and intestine [1]. The main turn in the use of zebrafish as an in vivo model for cancer research was the possibility to manipulate the zebrafish genome to recreate driver mutations present in human tumors. Nowadays, numerous transgenic cancer model lines are used for understanding the molecular mechanisms involved in tumor formation and progression [4]. The recent introduction of the CRISPR/Cas9 technology [5] has enabled a

Tumor Vascularization
DOI: https://doi.org/10.1016/B978-0-12-819494-2.00008-0

more reliable and fast gateway to transgene delivery and genetic engineering when compared to previous approaches based on TALEN or zinc finger nucleases technologies.

Information about cancer onset, growth, and metastatization can be obtained from zebrafish at different stages of development. In fish adults, cancer can be generated through the creation of a transgenic model line or via the grafting of cancer cells, even though limitations related to immunologic rejection of the graft may exist. Restrictions related to tumor imaging in adult animals have been partially solved with the introduction of the transparent zebrafish lines casper and crystal [6]. Transgenic lines and tumor cell grafting strategies can also be applied to juvenile fishes [between 10 and 45 days postfertilization (dpf)] to better understand early tumorigenesis. Juvenile fish remain fairly translucent, making them suitable for in vivo confocal microscopy to study stroma–tumor interaction [4].

When compared to adult and juvenile animals, the zebrafish embryo offers unique features that make it a unique model for the study of cancer. Fish fecundation is external and a high number of eggs can be obtained from a single breeding, allowing the manipulation of numerous embryos during their rapid development. Embryos do not require feeding during the first 5 dpf and can be maintained in small petri dishes at 28°C. They are nearly optically transparent during the first 24 hours postfertilization (hpf), and pigmentation can be abolished by adding *N*-phenylthiourea to the fish water [7], facilitating their observation by various microscopy techniques (see below).

Tumors induce sprouting angiogenesis as a strategy for survival and invasion [8]. Notably, the molecular mechanisms of tumor angiogenesis in zebrafish are similar to those observed in humans [9]. Since zebrafish embryo has the unique capacity to survive without a functional or fully formed blood circulation [10], interventions that disrupt blood vessels can be performed without being lethal to the embryo. The possibility to knockdown the expression of a given gene by injection of specific oligonucleotide morpholinos [11] allows the study of genes involved in pivotal biological processes, including vascular development. Moreover, embryos lack an adaptive immune system and, for this reason, murine and human tumor cells can be injected in internal structures, cavities, and developing organs with no rejection.

In this chapter, we will describe the use of zebrafish embryo as a tool for the study of tumor angiogenesis by focusing on the development of the vascular districts usually examined in these studies, the techniques to alter the physiological angiogenesis process and the interplay between tumor and host microenvironment. Lastly, we will describe the imaging techniques and equipment currently available to visualize blood vessels in zebrafish embryos.

Modeling angiogenesis in zebrafish embryo

Vascular endothelial cells (ECs) differentiate from mesoderm-derived precursors. Only recently, Npas4l has been identified as a fundamental transcriptional factor that drives the mesodermal differentiation into endothelial and hematopoietic progenitors. The vasculogenic process in zebrafish embryo arises around 14 hpf when endothelial precursors migrate from the lateral plate mesoderm toward the midline to form a vascular cord located under the hypochord. At 17 hpf, the arteriovenous differentiation occurs and ECs begin to express specific markers such as *ephrin-b2a* for ECs that will give rise to the dorsal aorta (DA) and *ephb4a* for ECs that will contribute to the posterior cardinal vein (PCV) and to the caudal vein. Arterial specification is guided by Sonic hedgehog and produced by the notochord that leads to activation of *vascular endothelial growth factor-a* (*vegfa*) expression in ventral somites. Interaction of Vegfa with the Vegf-receptor-2, named Kdrl in zebrafish, leads to the activation of Notch signaling via the transcription factors Mef2 and SoxF. At 21 hpf, the cells located in the ventral part of the vascular cord accumulate ventrally to form the DA, a cord-like structure that will reorganize in a patent vessel at around 24 hpf. At this time point, venous ECs also aggregate and organize to form the venous vasculature under the already formed DA [12,13].

Angiogenesis of intersegmental vessels

In zebrafish, the first angiogenetic process easily visualized is the formation of the intersegmental vessels (ISVs) that are part of the trunk vasculature and run between the somites (Fig. 8.1A–D). This two-step process begins at around 22 hpf, when some ECs sprout from the DA and form the segmental arteries (SA). In the sprouting process, the basal cells of the sprout, named stalk cells, have proliferative activity, whereas the leading cells, named tip cells, sense the angiogenic stimuli and protrude filopodia to spread (Fig. 8.1E). At the beginning of the ISV formation process, each sprout is made by one tip cell and two to three basal stalk cells [13]. Tip and stalk cells can be distinguished from nonsprouting ECs for the expression of the H2.0-like Homeobox-1 (hlx1) gene that represents a marker for angiogenic ECs [14]. ISV sprouts grow dorsally, reach the dorsal neural tube, the basic structure of the SAs is shaped and the connection of SAs with the anterior and posterior sprouted vessels leads to the formation of the dorsal longitudinal anastomotic vessels. At 32 hpf, a second wave of vascularization occurs when ECs sprout from the PCV and connect to SAs, thus forming the segmental veins [13]. The players that orchestrate this physiological angiogenesis

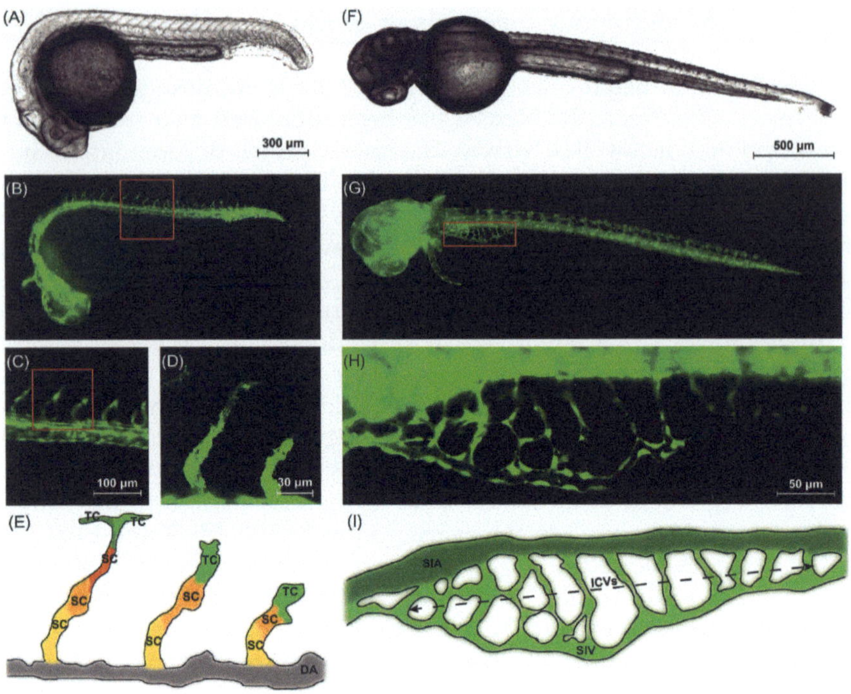

FIGURE 8.1 Bright field (A, F) and epifluorescent (B, G) representative pictures of whole-mount tg(*fli1a*:EGFP) zebrafish embryos at 28 hpf (A, B) and 48 hpf (F, G). Magnifications of the highlighted areas show detailed view of ISV (C, D) and SIVP (H) regions. (B, C, G, and H) Conventional epifluorescence microscopy. (D) Confocal microscopy. (E, I) Schematic representation of ISVs and SIVP, respectively. *DA*, dorsal aorta; *ICVs*, interconnecting vessels; *SC*, stalk cell; *SIA*, subintestinal artery; *SIV*, subintestinal vein; *TC*, tip cell. Source: *Chiara Tobia and Andrea Barbieri.*

process are numerous and their identification can be achieved by interfering with the normal ISV sprouting by genetic manipulation or drug administration.

As mentioned earlier, zebrafish embryo is suitable for genetic knockdown through the injection of specific antisense morpholinos at one-cell stage. These morpholino-modified oligonucleotides are designed to bind the ATG start codon or a splicing site of the RNA of interest, thus preventing, de facto, the protein translation [11]. Using this approach, the role of the *Vegf* gene family in the vascular development has been extensively described, the loss of function of Vegfa causing an impairment of ISV formation in zebrafish embryo [15]. More recently, a reverse genetic screen has been performed to identify new targets to modulate angiogenesis. Firstly, transcript profiling in mice lead to the identification of 150 potentially druggable microvessel-enriched gene products; among them, 62 orthologue genes were identified in zebrafish that were suitable for genetic knockdown by specific morpholino injection. When embryos were analyzed for vascular defects, downregulation of 16 genes caused significant vessel alterations in response to morpholino injection [16].

In addition to a morpholino approach, the possibility exists to create mutant zebrafish lines to study the involvement of a gene of interest in sprouting angiogenesis. For instance, Lawson and Weinstein [17] identified vascular mutants in zebrafish through a forward genetics screening approach. In particular, these authors were able to determine an allelic series of mutations in phospholipase c gamma 1 and studied the role of this gene in vascular development through the creation of mutant models.

Once the players involved in the angiogenesis process have been identified, they can become the targets for new antiangiogenic drugs. For instance, PTK787/ZK222584, an anilinophthalazine compound able to inhibit VEGF receptors with high affinity, suppresses vessel formation when administered at early stages of development. This phenotype recapitulates the phenotype observed in Vegfa morphants. However, while morpholino injection is performed at one-cell stage, chemical compounds can be administered at different stages of embryonic development, leading to different biological effects. For instance, PTK787/ZK222584 administration to 24 hpf embryos does not affect the entire vasculature whereas ISVs fail to form [18].

Zebrafish embryo is particularly suitable for high-throughput drug screenings. The assay can be carried out in 96-well plates by incubating the embryos with different chemicals at different concentrations, and the resulting phenotypes are then analyzed in vivo. For instance, taking advantage of zebrafish transgenic lines to visualize blood vessels under an epifluorescence microscope, an automated screening assay has been set up in which embryos treated with different compounds from the Sigma Life Science's Library of Pharmacologically Active Compounds (LOPAC1280) were analyzed for angiogenic defects. As a result, a new antiangiogenic compound, named IRO, was identified [19].

Subintestinal venous plexus development

The subintestinal plexus represents another site for new vessel formation in the trunk of the zebrafish embryo (Fig. 8.1F–H). It develops bilaterally on the yolk ball and provides blood to the gut, liver, and pancreas. Due to its venous origin [20,21], it is also named subintestinal venous plexus (SIVP) and it is formed by the subintestinal vein (SIV), the supraintestinal artery (SIA), and the interconnecting vessels (ICVs)

(Fig. 8.1I). The development of SIVP has been recently described in depth [22–24] and can be considered as a novel mechanism of vessel formation, distinct from the vasculogenic and angiogenic processes. Starting from 28 hpf, specialized angioblasts, already present in the ventral side of the PCV and in anterior PCV, migrate anterior-to-posterior and coalesce to form the SIV and subsequently the SIA. In situ hybridization (ISH) experiments showed that vegfaa is expressed by the endoderm surrounding the yolk between 48 and 72 hpf, and both vegfaa and vegfab are expressed by podocytes of pronephric ducts. Embryo treatment with chemical inhibitors, morpholinos or mutants targeting members of the Vegf/Vegfr family have demonstrated the role of this signaling pathway in driving angioblast migration from PCV, with vegfaa and its receptors acting probably as master regulator genes [24]. Additionally, bmp4 is expressed in the gut and BMP signaling is important for the ventral migration of SIV. During development, SIVP undergoes several modifications, including cell proliferation, compartment formation, remodeling, and pruning. At the end of the process, at about 4 dpf, it consists of a well-organized basket of vessels with a superficial and an inner plexus [23]. Interestingly, SIVs and ICVs express the specific lymph-venous marker lyve1, whereas SIAs acquire an arterial identity [22]. As circulation initiates, blood flow takes place with a rostral to caudal direction in the SIA and with an opposite direction in the SIV, whereas a dorsoventral direction is established in ICVs.

Subintestinal venous plexus and soluble angiogenic stimuli

Due to its anatomical position, the developing SIVP region is highly accessible and suitable for testing pro- and antiangiogenic stimuli [25,26]. To this respect, a zebrafish yolk membrane (ZFYM) assay has been developed on the basis of the injection of proangiogenic factors in 48 hpf embryos [27]. Briefly, embryos are anesthetized, placed on an agarose-coated plate, and injected in the perivitelline space with a proangiogenic mediator using a borosilicate capillary and a microinjector. The angiogenic response, consisting of ectopic vessels sprouting from the SIV, is evaluated 24 hours thereafter by endogenous alkaline phosphatase (AP) staining. Moreover, new vessel growth can be monitored in vivo and in a time-controlled manner using endothelial-specific transgenic reporter lines (see "Vessel imaging in zebrafish embryos" Section for a detailed description of techniques for the visualization of vessels in zebrafish). The ZFYM assay can be used to evaluate the proangiogenic effect of different families of molecules, including VEGFs, fibroblast growth factor-2 (FGF2), and proinflammatory cytokines and chemokines produced by tumor cells. Moreover, the ZFYM assay can be adapted to test conditioned media derived from cancer cell lines or stromal/inflammatory cell components cultured under different experimental conditions. Quantitative data can be obtained by evaluating the area of neovascularization or the vessel-specific fluorescence intensity, by counting the number of newly formed vessels and branching points or by assigning a score by trained observer(s) based on the extent of the response.

Subintestinal venous plexus and tumor-induced angiogenesis

To investigate the angiogenic response induced by cancer cells, a zebrafish embryo tumor/xenograft model has been established [28]. In this assay, a limited number of cancer cells suspended in Matrigel or PBS are injected in the perivitelline space of 48 hpf embryos. New blood vessel formation originating from the SIVs is evaluated starting from 24 hours postinjection (Figs. 8.2 and 8.3). With this technique, a single operator can inject 25–30 embryos in 1 hour. The large sample size increases significantly the statistical strength of each experiment.

The zebrafish embryo tumor/xenograft assay can be applied to evaluate in a quantitative manner the angiogenic potential of genetically modified cancer cells, overexpressing or silenced for the gene of interest. For instance, coadministration of the soluble pattern recognition receptor long-pentraxin 3 [28] or overexpression of miRNA-492 [29] suppresses the angiogenic activity exerted in this assay by murine FGF2-overexpressing murine aortic ECs and human DU-145 prostate cancer cells, respectively. Conversely, LIMK1/2 knockdown reduces the angiogenic ability of pancreatic cancer cell grafts when compared to the parental cell line [30].

On the other hand, the possibility to modulate the host vasculature via chemical inhibitors, morpholino knockdown, CRISPR/Cas9-mediated knockout or knock-in, may help to elucidate the role of specific genes of the host during the angiogenic process triggered by cancer cell implantation [31].

As mentioned above, tissue microenvironment conditioned by tumor cell plays a pivotal role during cancer progression and new vessel formation. Host myeloid cells support the neovascularization of localized tumor mass generated in zebrafish embryo after injection of cancer cells into the bloodstream [32]. Moreover, coinjection in the perivitelline space of cancer cells and tumor-associated macrophages (TAMs) enhances tumor dissemination [33]. Thus the zebrafish embryo tumor/xenograft assay may be used to perform injection of cells isolated from the tumor microenvironment, such as cancer-associated fibroblasts, TAMs (M1- vs M2-polarized macrophages), or myeloid cells, to evaluate

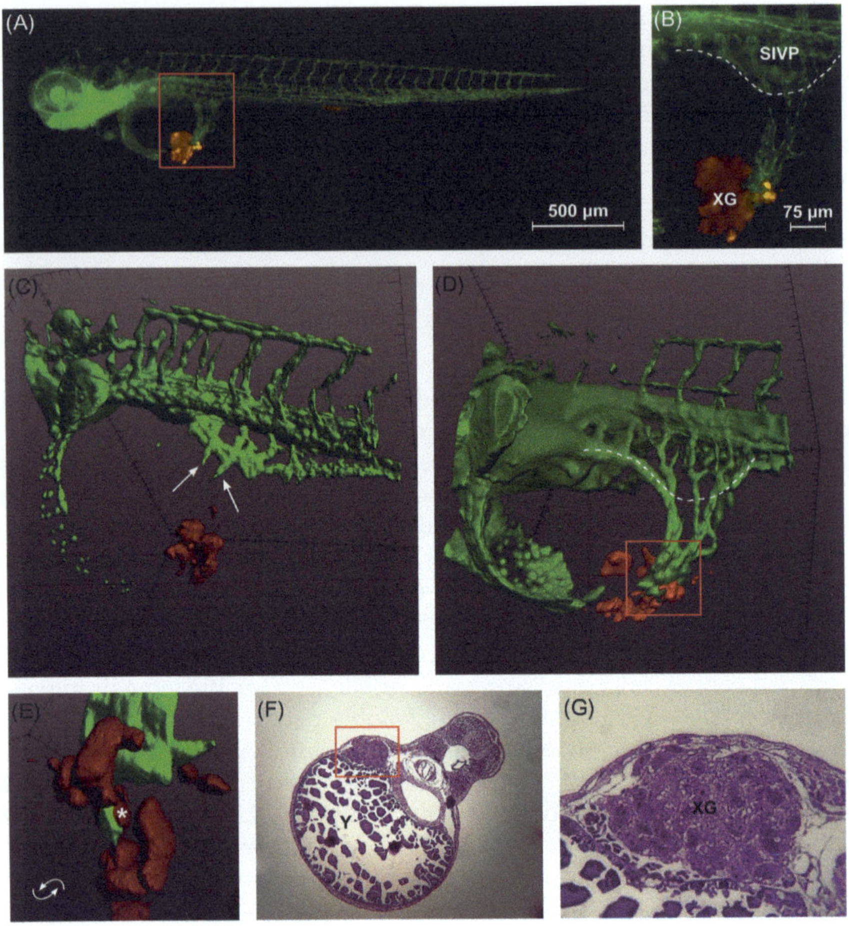

FIGURE 8.2 Xenotransplantation of tumor cells (*red*) within the perivitelline space induces an angiogenetic response in the SIVP of zebrafish embryos. (A) Representative whole-mount fluorescence image of a 72-hpf tg(*fli1a*:EGFP) embryo at 24 hours postinjection (hpi) and (B) magnification of the highlighted region. (C, D, and G) High-resolution 3D reconstruction by light sheet microscopy allows the identification of early vessel sprouts (*arrows* in C, acquired 3 hpi) and the crosstalk between the tumor cells and the host vasculature (D, G, acquired 24 hpi). *Striped lines* in (B) and (D) depict the physiological shape of the SIVP, heavily disrupted by the presence of the xenograft (*XG*). (G) The magnification of the highlighted region in (D), acquired at a different angle. The *asterisk* indicates a close contact between a tumor cells and the newly formed vessel. (E) Semi fine cross section of a zebrafish embryo showing the tumor mass and (F) magnification of the highlighted area. Toluidine blue staining. *EGFP*, Enhanced green fluorescent protein; *XG*, xenograft; *Y*, yolk. Source: *Chiara Tobia and Andrea Barbieri. Images in (F) and (G) were kindly provided by Prof. R. Ribatti, University of Bari, Italy.*

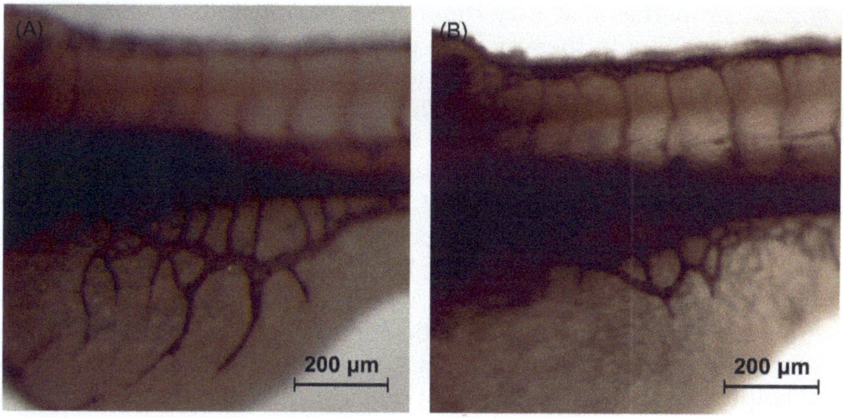

FIGURE 8.3 Alkaline phosphatase staining of zebrafish embryos injected with a proangiogenic stimulus in absence (A) and in presence (B) of an angiogenesis inhibitor. Source: *Jessica Guerra.*

the contribution of a specific environmental component to the development of new vessels in a nontumorigenic milieu. In addition, the zebrafish embryo tumor/xenograft assay has been employed to study the impact of hypoxia on tumor dissemination and neovascularization [34]. Several methods to induce hypoxia in both embryos and adults have been proposed in zebrafish [35]. When combined with tumor/stromal cell xenotransplantation, they may help to elucidate the contribution of tumor microenvironment to cancer progression and response to therapy.

The identification of new pharmacological compounds that target the molecular and cellular pathways involved in the angiogenetic process remains one of the goals of anticancer therapies [36]. Libraries of small chemicals or natural compounds can be added to fish water to test their effect on SIV development in zebrafish embryos [37]. The same approach can be used to investigate the potential inhibitory activity of a molecule or a library of molecules on soluble factor— or cancer cell—induced angiogenesis (see Ref. [36]). In addition, bloodstream injection represents an alternative method to deliver small drugs, nanocarriers, or functionalized nanoparticles ([38] and references therein) to target the tumor mass and newly formed vessels in order to prevent cell growth and metastatization. Of note, by combining the zebrafish embryo tumor/xenograft assay and the delivery of nanocarriers, zebrafish embryo can represent an ideal platform to obtain important in vivo information in nanomedicine, including nanoparticle toxicity, biodistribution, stability, effective targeting, and ability to exert pharmacological functions.

Immortalized cancer cell lines differ significantly from the original cancer cell population. Moreover, each tumor as unique genetic and biological properties that may differ from patient to patient, leading to different responses to therapy. The possibility to adjust the therapeutic plan for a single patient in order to choose the best pharmacological approach is becoming a real possibility. Thus in the frame of a personalized medicine approach to cancer therapy, patient-derived xenografts have been set up in zebrafish by several authors. To this respect, the zebrafish embryo tumor/xenograft assay represents a powerful and promising tool given the small amount of cells required for each transplantation and the rapid angiogenic response of the host after cell injection [39,40].

Angiogenesis favors the metastatic spreading of cancer cells into the bloodstream. Tumor xenotransplantation models in zebrafish embryos may help to elucidate the different steps of hematic dissemination of tumor cells via in vivo high-resolution imaging techniques (see "Imaging devices" Section).

Vessel imaging in zebrafish embryos

As mentioned above, zebrafish embryos provide several advantages over other animal models that make them particularly suitable for investigation in tumor angiogenesis. In the following paragraphs, we will review the most utilized imaging techniques and provide a brief overview of the imaging devices employed in the visualization of tumor vessels in both fixed and live zebrafish embryos.

Fixed embryo imaging

Visualization of blood vessel in fixed, nonliving embryos can be achieved by two different strategies: (1) ISH using antisense probes against vascular-restricted mRNA transcripts and (2) endogenous AP staining.

In situ hybridization

This well-established method takes advantage of labeled antisense RNA probes directed against vascular-restricted transcripts to pinpoint their spatial-temporal expression throughout the embryo [41]. Jung et al. [42] have recently reported a comprehensive list of vascular marker genes used in zebrafish research. Among them, *fli1a, kdrl, tie2,* and *cdh5* represent the most utilized markers to identify tumor angiogenesis. Obviously, other tissue-specific markers can be employed to further characterize the tumor microenvironment.

fli1a encodes for a DNA binding protein initially expressed in the posterior lateral mesoderm in hematopoietic and endothelial precursors. Then, *fli1a* expression becomes almost completely restricted to ECs with two minor exceptions, namely cranial neural crest derivatives and a small subset of circulating myeloid cells.

kdrl (*kdr-like*) has no an orthologous counterpart in mammals. It encodes for a fourth tyrosine-kinase Vegf receptor alongside Flt-1 (Vegfr1), Flk-1/Kdr (Vegfr2), and Flt-4 (Vegfr3). Kdrl is particularly involved in Vegf-guided sprouting of new blood vessels during angiogenesis and undergoes three different waves of expression. It is first expressed in the lateral mesoderm along with *flt-4*; then its expression is restricted to angioblasts and endothelium; finally, it is highly expressed in the DA at later stages of embryonic development (whereas *flt-4* expression is restricted to the PCV).

tie2 and *cdh5* are the zebrafish orthologous counterparts of *Angiopoietin-1 receptor* and *vascular-endothelial cadherin*, respectively. Their expression is typically vascular.

Even though various ISH detection methods are available, indirect methods are usually preferred over direct methods as they provide a strong signal amplification, which is mandatory to pinpoint even poorly expressed transcripts and to reduce the timing for signal detection.

Briefly, fixed zebrafish embryos are incubated with a biotin-, digoxigenin-, or fluorescein-labeled antisense probe that hybridizes to the specific transcript. Then, following immunodecoration with antilabel, AP-conjugated antibodies, colorimetric signal detection is performed with a proper substrate system (e.g., 5-bromo-4-chloro-3-indolyl phosphate/NitroBlue tetrazolium chloride pair, Fast Red, Fast Blue, or BM-Purple). Two or more alternatively labeled probes can be detected simultaneously performing sequential incubations with different antibody/chromogen pairs [43].

Along with the colorimetric AP-based ISH, other ISH visualization approaches are nowadays available. The Fast Red/Blue pair, for example, can also be fluorescently visualized [43]. The tyramide signal amplification system [44] employs horseradish peroxidase–conjugated antibodies and tyramide-conjugated molecules to increase the sensitivity and enable the use of fluorophores. Finally, the RNA-Scope technology [45] uses ad hoc synthesized probe/amplifier sets to build a scaffold for fluorescent labels, greatly boosting the sensitivity and allowing easy multiplexing.

Alkaline phosphatase staining

Zebrafish blood vessels possess endogenous AP activity. Negligible at 24 hpf, AP activity increases at 2 dpf and peaks at 3 dpf. It is worth noting that fixation does not hinder AP activity. This implies the use of endogenous AP inhibitors, such as tetramisole hydrochloride, when performing a colorimetric ISH as described above. On the other hand, endogenous AP activity can be exploited to carry out a fast and easy visualization of vasculature in 3 dpf embryos, even though at a lower resolution due to a higher background (Fig. 8.3). Despite these limitations, AP staining is by far the easiest and fastest method to visualize SIV sprouts in fixed embryos [26].

Live embryo imaging

The abovementioned approaches allow a comprehensive analysis of the angiogenic process at a fixed time of the experimental protocol. However, the best way to capitalize on the use of zebrafish embryos in tumor angiogenesis investigation is to perform vital imaging of the embryo vasculature, which is favored by the availability of numerous fluorescent protein reporter zebrafish lines. Differently from nonvital blood vessel imaging, these lines allow a time-lapse oriented, high-resolution imaging of the blood vessels without interfering with embryo development.

Fluorescent reporter lines

An extensive list of the vascular reporter zebrafish lines nowadays available has been reported [42]. Among them, *fli1a:EGFP* [17] and *kdrl:EGFP* [46] lines expressing cytoplasmic enhanced green fluorescent protein (EGFP) under the *fli1a* or *kdrl* promoter, respectively, represent by far the most used zebrafish models in tumor angiogenesis investigation. The same promoters have also been used to generate additional transgenic lines, in which fluorescent proteins other than EGFP are expressed in the vascular district (e.g., DsRed, Cherry, RFP, YFP, Kaede), or where the localization of the fluorescent protein is nuclear (nEGFP) or membranous (EGFP-cdc42wt). In addition, the implementation of the GAL4–UAS system within the zebrafish model has further enriched the plasticity of zebrafish imaging, and the recent introduction of the CRISPR/Cas9 technique has laid the basis for a faster construction of new reporter as well as mutant zebrafish lines.

Given the versatility of the zebrafish embryo and the broad collection of reporter lines and models, the whole neoangiogenesis process can be thoroughly investigated in vivo while the tumor is still progressing. In addition, fluorescent cell trackers enable the evaluation of the crosstalk between labeled tumor cell grafts and the preexistent vasculature, as well as the dissection of the dissemination process throughout the whole embryo up to single-cell metastases. This approach, however, is not suited for long-term experiments as the fluorescent cell tracker will increasingly dilute and diffuse within the embryo. Thus fluorescent protein transduced cells, if available, are undoubtedly the best choice for tumor grafting in zebrafish. Last, dedicated imaging techniques and devices, such as light sheet microscopy, allow a three-dimensional (3D) fluorescent reconstruction of the tumor–vessel interface at unprecedented speed, the whole embryo being imaged and reconstructed in minutes.

Microangiography

Visualization of patent blood vessels supporting an active circulation can be carried out also in nonfluorescent fish lines through a microangiographic approach [47]. Proper fluorescent dyes, such as rhodamine-labeled dextran or quantum dots, are injected directly in the bloodstream. A precise reconstruction of the vasculature tree is retrieved by 3D imaging techniques such as confocal, multiphoton, or light sheet microscopy. Two to three dpf embryos represent the ideal candidate for microangiography as injections become increasingly more difficult as the embryo grows. In this frame, typical injection sites are either the *sinus venosus*, the PCV, or the duct of Cuvier. The major flaws of this approach reside in the impossibility to perform repeated injections in the same embryo as well as the lack of information deriving from nonpervious vessels or from embryos with impaired circulation.

Imaging devices

Taking into account the central role played by image acquisition, and processing steps in tumor angiogenesis investigations, adequate devices, and protocols are indeed required in order to capitalize on the zebrafish model.

Epifluorescence stereomicroscopes equipped with digital cameras represent by far the core devices every laboratory should use in order to carry out most of the routine analysis. Thanks to their relative low price and plenty of room for manipulation, they are indeed the best choice in terms of acquisition speed and ease of use. Depending on the selected hardware and the software bundle, it is possible to gain access to several enhanced features, such as time-lapse and Z-stack acquisition, 3D reconstruction, tile scanning for large samples, structured illumination and automated or semiautomated deconvolution algorithms for crisper images. For simple observation and brief acquisitions, embryos are usually placed in Petri dishes filled with PBS + Tween20 (fixed embryos) or with a 0.016% tricaine solution in embryo water as anesthetizing agent (living embryos). To keep the correct orientation and reduce tilting due to the presence of the yolk ball, small holes or trenches can be carved into an agarose coating to stabilize the embryos. Alternatively, embryos can be soaked into a high-density medium, such as methylcellulose and glycerol, the latter being suitable only for fixed embryos. For longer acquisition intervals of living embryos, erratic movements should be totally prevented: for this reason, embryos are usually embedded in low-melting agarose to hold the orientation and soaked in the tricaine solution to maintain the sedation throughout the whole experiment. However, the embryo should be allowed to grow normally and undergo all the required morphogenic rearrangements (see Ref. [42] for further details).

Confocal and multiphoton microscopes address two of the major limitations related to conventional fluorescence microscopes, namely low magnifying power and poor vertical resolution, at the cost of a strongly reduced sample acquisition rate. Both approaches take advantage of focused laser beams to stimulate the fluorescence reporters within the embryos whereas confocality is achieved in two different ways. Laser-scanning and spinning disc confocal microscopes stimulate an hourglass-shaped tissue volume and employ pinholes to eliminate the signal deriving from out-of-focus planes. At variance, the optical sectioning of multiphoton microscopes is an intrinsic feature of the device, as the photon density required for the fluorophore stimulation is present only within the in-focus plane. This explains why the use of a multiphoton device is strongly recommended for long-lasting imaging experiments, as it reduces the tissue damage due to frequent and prolonged laser radiation as well as improves the resolution of deep structures, increasing the outcome of the 3D reconstruction. Due to the long time required for the acquisition, living samples are usually embedded in low-melting agarose while fixed embryos can also be flat-mounted on glass slides.

Light sheet microscopy is the state-of-art technology for zebrafish imaging, representing the best choice for fluorescence-based morphological analyses. Even though it falls behind the confocal approach in the magnification power, high-contrast images are generated in a reasonably short amount of time thanks to an outstanding optical sectioning capability. In contrast to confocal and multiphoton microscopy, light sheet microscopy illuminates the sample orthogonally to the observer by means of a laser source focused by a cylindrical lens into a thin light blade. This way, the sample is analyzed plane-by-plane rather than "scanned" point-by-point, increasing significantly the acquisition performance. Zebrafish embryos are usually anesthetized and mounted in a low-melting agarose cylinder, which is extruded from a glass capillary. A motorized manipulator is then employed to perform 4D movements (three axes *plus* rotation) to finely adjust the embryo position inside the tricaine-filled imaging chamber. This setup is very convenient for time-lapse acquisitions, as it ensures embryo immobilization and reduces photo-damage to the minimum when compared to the confocal approach. Moreover, drugs can be easily administered to the embryo inside the imaging chamber.

Altogether, experimental evidence highlights the zebrafish embryo as a valid animal model, complementary to already established rodent models, to obtain robust evidence to support in vitro and in vivo translational research in tumor angiogenesis.

Acknowledgment

This work was supported by grant IG 18943 from Associazione Italiana per la Ricerca sul Cancro (AIRC) to MP.

References

[1] Stern HM, Zon LI. Cancer genetics and drug discovery in the zebrafish. Nat Rev Cancer 2003;3(7):533–9.

[2] Howe K, et al. The zebrafish reference genome sequence and its relationship to the human genome. Nature 2013;496(7446):498–503.

[3] Barriuso J, Nagaraju R, Hurlstone A. Zebrafish: a new companion for translational research in oncology. Clin Cancer Res 2015;21(5):969–75.

[4] White R, Rose K, Zon L. Zebrafish cancer: the state of the art and the path forward. Nat Rev Cancer 2013;13(9):624–36.

[5] Irion U, Krauss J, Nusslein-Volhard C. Precise and efficient genome editing in zebrafish using the CRISPR/Cas9 system. Development 2014;141(24):4827–30.

[6] Antinucci P, Hindges R. A crystal-clear zebrafish for in vivo imaging. Sci Rep 2016;6:29490.

[7] Karlsson J, von Hofsten J, Olsson PE. Generating transparent zebrafish: a refined method to improve detection of gene expression during embryonic development. Mar Biotechnol (NY) 2001;3(6):522–7.

[8] Carmeliet P, Jain RK. Angiogenesis in cancer and other diseases. Nature 2000;407 (6801):249–57.

[9] Tobia C, De Sena G, Presta M. Zebrafish embryo, a tool to study tumor angiogenesis. Int J Dev Biol 2011;55(4–5):505–9.

[10] Sehnert AJ, et al. Cardiac troponin T is essential in sarcomere assembly and cardiac contractility. Nat Genet 2002;31(1):106–10.

[11] Nasevicius A, Ekker SC. Effective targeted gene 'knockdown' in zebrafish. Nat Genet 2000;26(2):216–20.

[12] Matsuoka RL, Stainier DYR. Recent insights into vascular development from studies in zebrafish. Curr Opin Hematol 2018;25(3):204–11.

[13] Ellertsdottir E, et al. Vascular morphogenesis in the zebrafish embryo. Dev Biol 2010;341(1):56–65.

[14] Herbert SP, Cheung JY, Stainier DY. Determination of endothelial stalk versus tip cell potential during angiogenesis by H2.0-like homeobox-1. Curr Biol 2012;22 (19):1789–94.

[15] Nasevicius A, Larson J, Ekker SC. Distinct requirements for zebrafish angiogenesis revealed by a VEGF-A morphant. Yeast 2000;17(4):294–301.

[16] Kalen M, et al. Combination of reverse and chemical genetic screens reveals angiogenesis inhibitors and targets. Chem Biol 2009;16(4):432–41.

[17] Lawson ND, Weinstein BM. In vivo imaging of embryonic vascular development using transgenic zebrafish. Dev Biol 2002;248(2):307–18.

[18] Chan J, et al. Dissection of angiogenic signaling in zebrafish using a chemical genetic approach. Cancer Cell 2002;1(3):257–67.
[19] Tran TC, et al. Automated, quantitative screening assay for antiangiogenic compounds using transgenic zebrafish. Cancer Res 2007;67(23):11386–92.
[20] Lenard A, et al. Endothelial cell self-fusion during vascular pruning. PLoS Biol 2015;13(4):e1002126.
[21] Nicenboim J, et al. Lymphatic vessels arise from specialized angioblasts within a venous niche. Nature 2015;522(7554):56–61.
[22] Hen G, et al. Venous-derived angioblasts generate organ-specific vessels during zebrafish embryonic development. Development 2015;142(24):4266–78.
[23] Goi M, Childs SJ. Patterning mechanisms of the sub-intestinal venous plexus in zebrafish. Dev Biol 2016;409(1):114–28.
[24] Koenig AL, et al. Vegfa signaling promotes zebrafish intestinal vasculature development through endothelial cell migration from the posterior cardinal vein. Dev Biol 2016;411(1):115–27.
[25] Raghunath M, et al. Pharmacologically induced angiogenesis in transgenic zebrafish. Biochem Biophys Res Commun 2009;378(4):766–71.
[26] Serbedzija GN, Flynn E, Willett CE. Zebrafish angiogenesis: a new model for drug screening. Angiogenesis 1999;3(4):353–9.
[27] Nicoli S, De Sena G, Presta M. Fibroblast growth factor 2-induced angiogenesis in zebrafish: the zebrafish yolk membrane (ZFYM) angiogenesis assay. J Cell Mol Med 2009;13(8B):2061–8.
[28] Nicoli S, Presta M. The zebrafish/tumor xenograft angiogenesis assay. Nat Protoc 2007;2(11):2918–23.
[29] Chiavacci E, et al. The zebrafish/tumor xenograft angiogenesis assay as a tool for screening anti-angiogenic miRNAs. Cytotechnology 2015;67(6):969–75.
[30] Vlecken DH, Bagowski CP. LIMK1 and LIMK2 are important for metastatic behavior and tumor cell-induced angiogenesis of pancreatic cancer cells. Zebrafish 2009;6(4):433–9.
[31] Nicoli S, et al. Mammalian tumor xenografts induce neovascularization in zebrafish embryos. Cancer Res 2007;67(7):2927–31.
[32] He S, et al. Neutrophil-mediated experimental metastasis is enhanced by VEGFR inhibition in a zebrafish xenograft model. J Pathol 2012;227(4):431–45.
[33] Wang J, et al. Novel mechanism of macrophage-mediated metastasis revealed in a zebrafish model of tumor development. Cancer Res 2015;75(2):306–15.
[34] Lee SLC, et al. Hypoxia-induced pathological angiogenesis mediates tumor cell dissemination, invasion, and metastasis in a zebrafish tumor model. Proc Natl Acad Sci USA 2009;106(46):19485–90.
[35] Elks PM, et al. Exploring the HIFs, buts and maybes of hypoxia signalling in disease: lessons from zebrafish models. Dis Model Mech 2015;8(11):1349–60.
[36] Santoro MM. Antiangiogenic cancer drug using the zebrafish model. Arterioscler Thromb Vasc Biol 2014;34(9):1846–53.
[37] Zhu XY, et al. Closantel suppresses angiogenesis and cancer growth in zebrafish models. Assay Drug Dev Technol 2016;14(5):282–90.
[38] Gutierrez-Lovera C, et al. The potential of zebrafish as a model organism for improving the translation of genetic anticancer nanomedicines. Genes (Basel) 2017;8(12).
[39] Wu JQ, et al. Patient-derived xenograft in zebrafish embryos: a new platform for translational research in gastric cancer. J Exp Clin Cancer Res 2017;36(1):160.
[40] Gaudenzi G, et al. Patient-derived xenograft in zebrafish embryos: a new platform for translational research in neuroendocrine tumors. Endocrine 2017;57(2):214–19.
[41] Thisse B, Thisse C. In situ hybridization on whole-mount zebrafish embryos and young larvae. Methods Mol Biol 2014;1211:53–67.

[42] Jung HM, et al. Imaging blood vessels and lymphatic vessels in the zebrafish. Methods Cell Biol 2016;133:69–103.
[43] Lauter G, Soll I, Hauptmann G. Two-color fluorescent in situ hybridization in the embryonic zebrafish brain using differential detection systems. BMC Dev Biol 2011;11:43.
[44] Bobrow MN, et al. Catalyzed reporter deposition, a novel method of signal amplification. Application to immunoassays. J Immunol Methods 1989;125(1–2):279–85.
[45] Wang F, et al. RNAscope: a novel in situ RNA analysis platform for formalin-fixed, paraffin-embedded tissues. J Mol Diagn 2012;14(1):22–9.
[46] Cross LM, et al. Rapid analysis of angiogenesis drugs in a live fluorescent zebrafish assay. Arterioscler Thromb Vasc Biol 2003;23(5):911–12.
[47] Schmitt CE, Holland MB, Jin SW. Visualizing vascular networks in zebrafish: an introduction to microangiography. Methods Mol Biol 2012;843:59–67.

CHAPTER

9

Clinical strategies to inhibit tumor vascularization

Adrian L. Harris

Molecular Oncology Laboratories, Oxford University, Department of Oncology, Weatherall Institute of Molecular Medicine, John Radcliffe Hospital, Oxford, United Kingdom

Introduction

The discovery of vascular endothelial growth factor (VEGF, also called vascular permeability factor) heralded years of research to develop drugs to block its binding to cognate receptors or inhibit downstream kinases. These resulted in literally hundreds of phase I, II, and III trials in most tumor types, and helped establish the role of each drug in selected tumor types [1–5]. The current drugs (Table 9.1) and putative modes of action and US approvals are listed (Table 9.2).

However, initial enthusiasm was based on the paradigm of tumor angiogenesis, the development of new blood vessels from preexisting vessels and indeed there was substantial evidence for this. An example was the demonstration of increased endothelial proliferation by BUdR labeling in patients with breast cancer [6]. We now know there are many other ways for tumors to maintain or develop a blood supply or maintain oxygenation (e.g., neuroglobin [7]) revealed in this volume.

Very few of the trials tested biomarkers that were able to predict benefit or response before treatment and none were developed into clinical practice. Additionally, trial design was often flawed with lack of suitable control groups and drugs that could not be maintained because of toxicity. Examples particularly involve the kinase inhibitors that needed drug holidays, that is, a week off therapy to be treated. This obviously could allow rebound and regrowth of tumor.

Tumor Vascularization
DOI: https://doi.org/10.1016/B978-0-12-819494-2.00009-2

TABLE 9.1 Current inhibitors.

Axitinib (Inlyta)
Bevacizumab (Avastin)
Cabozantinib (Cometriq)
Everolimus (Afinitor)
Lenalidomide (Revlimid)
Lenvatinib mesylate (Lenvima)
Pazopanib (Votrient)
Ramucirumab (Cyramza)
Regorafenib (Stivarga)
Sorafenib (Nexavar)
Sunitinib (Sutent)
Thalidomide (Synovir, Thalomid)
Vandetanib (Caprelsa)
Ziv-aflibercept (Zaltrap)

From https://www.cancer.gov/about-cancer/treatment/types/immunotherapy/angiogenesis-inhibitors-fact-sheet.

Many different approaches were tried, including use with neoadjuvant chemotherapy, maintenance therapy after discontinuing chemotherapy and adjuvant therapy. It had been proposed that escape from tumor dormancy in some cases was related to vascularization of micrometastases. However, no adjuvant trial blocking angiogenesis has been successful so far [8]. This suggests that the paradigm is incorrect. A current hypothesis relates to more immunological mechanisms slow turnover populations.

A new insight into how angiogenic drugs may work was provided by Rakesh Jain's hypothesis of vascular normalization [9–11] (Fig. 9.1). In this scenario, the aberrant VEGF-dependent vessels reform better organized and perfused structures, which can also enhance delivery of other drugs. This was in contrast to the reduction in blood vessels and concomitant increase in hypoxia commonly seen in animal and other clinical studies.

Although initially controversial, animal and clinical studies in glioblastomas showed microvessel perfusion was increased by antiangiogenic therapy. This effectwas a marker of clinical benefit, as evidenced by improved outcomes for those patients with normalized and provided a new paradigm for therapy.

TABLE 9.2 Mechanisms and indications.

	US FDA approval year	EMA approval year	Comments
Bevacizumab			
Metastatic colorectal cancer	2004 (first line); 2006 (second line)	2005 (first and second line)	With chemotherapy, first and second line
Nonsmall-cell lung cancer	2006	2007	With chemotherapy, first line
Renal cell carcinoma	2009	2005	With interferon
Ovarian cancer	2014 for platinum-resistant recurrent ovarian cancer; not approved for first-line or platinum-sensitive recurrent ovarian cancer	2012 for first-line and platinum-sensitive recurrent ovarian cancer; 2014 for platinum-resistant recurrent ovarian cancer	With chemotherapy
Breast cancer	Withdrawn	2009	With chemotherapy
Regorafenib			
Refractory metastatic colorectal cancer	2013	2013	Single drug, refractory
Ramucirumab			
Gastric or gastroesophageal junction cancers	2014	2014	Refractory with or without chemotherapy
Nonsmall-cell lung cancer	2014	2014	Refractory with chemotherapy
Metastatic colorectal cancer	2015	Not yet approved	Refractory with chemotherapy
Sorafenib			
Hepatocellular carcinoma	2007	2006	Single drug, first line
Renal cell carcinoma	2005	2006	Single drug, first line
Thyroid cancer (differentiated)	2013	2014	Refractory to radioactive iodine
Sunitinib			
Renal cell carcinoma	2006	2007	Single drug, first line
Pancreatic neuroendocrine tumors	2011	2010	Single drug, progressive well differentiated pancreatic neuroendocrine tumors

(Continued)

TABLE 9.2 (Continued)

	US FDA approval year	EMA approval year	Comments
Pazopanib			
Renal cell carcinoma	2009	2009	Single drug, first line
Soft tissue sarcoma	2012	2012	Single drug
Axitinib			
Renal cell carcinoma	2012	2012	Single drug, second line
Vandetanib			
Medullary carcinoma of thyroid	2011	2012	Unresectable, locally advanced, or metastatic medullary thyroid cancer
Lenvatinib			
Thyroid cancer	2015	2015	Locally recurrent or metastatic, progressive, radioactive iodine-refractory differentiated thyroid cancer
Nintedanib			
Nonsmall-cell lung cancer	Not licensed for lung cancer	2014	Locally advanced, metastatic or second line nonsmall-cell lung cancer
Aflibercept			
Colorectal cancer	2012	2013	Second line

Notes: Several multikinase inhibitors are approved for use in patients with gastrointestinal stromal tumors, but we attribute this activity to nonangiogenic signaling (e.g., KIT inhibition). Other drugs are hypothesized to have antiangiogenic activity, and drugs not primarily developed as antiangiogenic drugs are not included in this table. No antiangiogenic drugs have been approved in the adjuvant setting for any tumor type. For more precise approval of the use of these drugs, see EMA and FDA.
EMA, European Medicines Agency; *FDA*, Food and Drug Administration.
Reproduced with permission from Jayson GC, Kerbel R, Ellis LM, Harris AL. Antiangiogenic therapy in oncology: current status and future directions. Lancet 2016;388(10043):518–29.

The problems are codelivery drugs in the concurrent time frame, the generalization of the effects and biomarkers to predict which patients would show the desired effect.

Overview of clinical studies

The table of approved drugs immediately shows that many common tumor types have an approved indication, for example, colon, ovary,

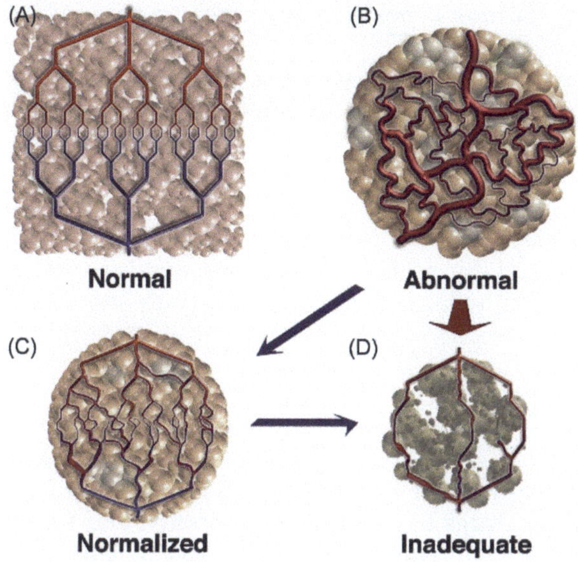

FIGURE 9.1 **Vascular normalization.** (A) Normal vasculature, composed of mature vessels and maintained by the perfect balance of pro- and antiangiogenic molecules, might not change during the course of antiangiogenic therapy. (B) Abnormal tumor vasculature, composed largely of immature vessels with increased permeability, vessel diameter, vessel length, vessel density, tortuosity and interstitial fluid pressure, compromises the delivery of therapeutics and nutrients. (C) Judiciously applied direct or indirect antiangiogenic therapies might prune immature vessels, leading to more normalized tumor vasculature. This network should be more efficient for the delivery of therapeutics and nutrients. (D) Rapid pruning of, or coagulation in, tumor vasculature might reduce the vasculature to the point that it is inadequate to support tumor growth and might lead to tumor dormancy. This is the ultimate goal of antiangiogenic/antivascular therapy. Source: *Reproduced with permission from Jain RK. Normalizing tumor vasculature with anti-angiogenic therapy: a new paradigm for combination therapy. Nat Med 2001;7(9):987–989.*

nonsmall-cell lung cancer (NSCLC), but other common tumors, such as breast and prostate cancer do not. Nearly all tumor types have been tested, so this shows a remarkable selectivity. The reasons for this are unknown, but the vascular beds of each organ are highly specialized and emerging single cell sequencing data, mainly from the mouse, show the heterogeneity at a transcriptome level [12,13].

Thus the gut endothelium has a major role in metabolite transport, the lung specialized in oxygen exchange. Each tissue will have different oxygen, nutrient level, and tumor forces. It was thought that all types of tumor would have common targets in their new vessels, but clinical trial data show that even if they express VEGFR, inhibiting them alone is insufficient in many tumor types. Each tumor type has been extensively well reviewed recently for trials and data will be summarized here.

Tumor types and response to therapy

There are marked differences in the response to anti-VEGF therapy, which do not seem explainable by the degree of vascularization of the tumor of simple measures of VEGFA content. This is a key area for investigation and necessary for further rational development of therapy. Recent advances in understanding vascularization as opposed to angiogenesis may help, only the latter requiring new vessels to grow, mainly stimulated by VEGF. The recent analyses of the metabolic pathways in endothelial cells are also likely to show tissue heterogeneity [14].

Lung cancer

Recent analyses and metaanalyses of all randomized trials of addition of antiangiogenic drugs to chemotherapy, in first and second lines, have shown no overall survival benefit [15]. Additionally, the drugs only seem to be of use in nonsquamous cancer as it can be associated with pulmonary hemorrhage in the latter [16]. In EGFR mutant lung cancer, there was no benefit. Progression-free survival was significantly benefitted as was response rate. Although a small significant increase in OS was seen in some subgroups, the recent recommendation by ASCO that a clinically meaningful HR for lung cancer was 0.8 would preclude those.

But recent striking results of combining chemotherapy, antiangiogenic therapy and the anti-PD-L1 antibody, atezolizumab, have provided major new impetus to antiangiogenic therapy. Median OS was increased by 4 months, regardless of molecular subtypes, but all were nonsquamous socinski. This combination is now licensed and a paradigm for other tumor types and antiangiogenic drugs in combination with immune checkpoint blockades (ICBs).

Ovarian cancer

Bevacizumab combined with chemotherapy was considered one of the major advances in gynecological cancer in 2018 [17]. Improvements in progression-free survival (PFS) but not OS were noted in several trials. These were for primary treatment after surgery, and also in relapsed platinum-resistant and platinum sensitive cases [18,19]. Although, as with many of these studies, subgroup analysis was performed and showed a possible group with increased OS, this was not based on any rationale.

Breast cancer

An early positive result for addition of bevacizumab to chemotherapy first line in metastatic breast cancer, using taxol [20], stimulated an extensive program of research into nearly every possible clinical scenario, from adjuvant, neoadjuvant, first and second lines in chemotherapy for metastatic cancer [21]. As for several other tumor types, there was a significant improvement in progression-free survival, but not in overall survival. The quality of life was similar, adverse effects were straightforward to manage, and there was no difference in effectiveness in prespecified subgroups. The effect was most pronounced in first line rather than second-line setting.

Food and Drug Administration (FDA) approved bevacizumab in 2008 for metastatic breast cancer, but it was a "conditional" accelerated approval, requiring further evidence and against the advice of their expert panel, which voted 5 to 4 against, subsequent to the larger number if trials and failure to show OS benefit.

Studies then focused on neoadjuvant therapy in HER2 positive and HER2 negative cases [22], five randomized trials included in the metaanalysis nearly 4500 patients. The addition of bevacizumab to chemotherapy significantly increased the pathological complete response (pCR) rate in both triple receptor negative (TRN) and estrogen receptor (ER) positive cases, 11% absolute for the former and 9% for the latter. There was no effect on disease-free survival, but this may be because an approximate 10% improvement in PCR is insufficient.

For HER2 positive cases, there were two trials and there was no difference in pCRrate [23].

As adjuvant therapy, after surgery, with chemotherapy and then maintained for 1 year, again, there was no benefit. Two trials were done, E5103 with 4994 patients in HER2 negative cases, and the BEATRICE trial, 2591 patients, TRN only [22].

In ER positive metastatic cases, the pooled results of the LEA and CALGS 40503 trials found that PFS was increased but with no effect on overall survival [24].

The FDA withdrew its approval in 2010. Thus, overall, in contrast to other major tumor types, there is no role for bevacizumab currently in breast cancer management. It does clearly have a significant effect in slowing tumor progression in recurrent disease and increased pCR; hence, a new wave of studies including immunotherapy is ongoing. This is particularly relevant in TRN breast cancer, a subgroup where immunotherapy is already approved and bevacizumab has a more marked effect [25,26]. In a landmark study, Schmid et al. showed in TRN advanced breast cancer there was a 10-month improved survival in those patients who were PD-L1 positive and received atezolizumab in addition to nab-paclitaxel.

Renal cancer

Antiangiogenic drugs were a mainstay of treatment of renal cancer, one of the most vascular of tumors [27–30]. Mutations in vHL led to upregulation of hypoxia-inducible factor 1 (HIF1) and HIF2 and hence high VEGF production besides many other cytokines and angiogenic factors [27]. However, recent developments in PD-L1 inhibitor therapies have moved immunotherapy to the frontline. Nevertheless, the combinations with the different antiangiogenic drugs and sequencing of drugs after first-line therapy will be informed by current extensive data from randomized trials. Several tyrosine kinase inhibitors (TKIs) have been approved in the first- and second-line settings. Rapid progress has been made in testing the most active of these TKIs, axitinib, lenvatinib, and tivozanib in combination with pembrolizumab, nivolumab, and atezolizumab, respectvely. Most recent data show the survival advantage of OCB alone versus TKI alone (nivolumab plus ipilimumab versus sunitinib [31] and atezolizumab plus bevacizumab versus sunitinib [32]). The mechanisms that allow for these substantial effects remain to be elucidated in the clinic.

Glioblastoma (high-grade glioma)

Eleven randomized trials, with 3743 participants, were analyzed by Ref [33]. As with other tumor types, there was no improvement in overall survival, although there was a highly significant improvement in progression-free survival. Additional clinical situations such as adjuvant, recurrent, with or without chemotherapy did not show a selective benefit for OS or PFS. The impact on quality of life was variable and an important consideration in this context. This therapy is not recommended in Brain Metastasis Guidelines of the US Congress of Neurological Surgeons [34].

However, a more mechanistic approach to understand if there are subgroups who do benefit in needed, for example, by analyzing differences in response such as blood flow improvements by vascular normalization, or induction of more severe hypoxia by the antiangiogenic effects.

Jain's work on vascular normalization exemplifies this, whereby they developed a vascular normalization index after a single dose of the VEGFR kinase inhibitor cerdiranib [35,36]. Changes in vascular permeability and flow and microvessel volume were measured by MRI. Thirty-one patients with recurrent glioma were studied. Two baseline scans were done, one day before and one day after cerdiranib. Those that showed the greatest drop in Ktrans had the best OS and PFS. Similarly, an increase in cerebral blood volume (CBV) was associated with an improved outcome. Such results of an early decrease of Ktrans and improved outcome have been reported for vatalanib in colorectal cancer and sunitinib in hepatoma [35]. The difference here is the increase in CBV, interpreted as vascular normalization, rather than the decrease that may be expected with decrease in blood vessels.

An extension of this into newly diagnosed glioblastome multiforme (GBM) treated with chemoradiation alone or in combination with cerdiranib [37]. Patients had incomplete tumor resections and temozolomide was used in standard dosage with radiation. All patients responded to cerdiranib with reductions in Ktrans and vessel size. However, 20 cases (50%) showed an increase in microvessel perfusion, occurring as early as day 1 after treatment. These patients had the better outcome [38].

A more complex effect was seen in recurrent glioblastoma using bevacizumab and MRI analysis [39].

Bevacizumab induced a strong antiangiogenic effect, with reduction in tumor oxygenation. But those with higher oxygenation had worse OS, suggesting that compensation mechanisms occurred in the more aggressive tumors. This study only used a single antiangiogenic agent and the window of normalization may be important for drug access. Additionally, radiotherapy may be more effective in the oxygenated tumors. This study emphasizes the additional utility of measuring oxygen metabolism, which will be extensively modified by baseline and targeted vasculature.

Colon cancer

Antiangiogenic drugs continue to have utility for this tumor type. In metastatic colon cancer, both as first- and second-line therapy [40]. Bevacizumab predominates, as for many other tumor types. Bowel perforation occurs as a more frequent problem, as would be expected. Only one study in second line was done up to 2009, and showed significant benefits in PFS and OS. A key issue was duration of bevacizumab therapy and whether it could continue to have a benefit beyond progression. It is possible that progression could represent resistance to the chemotherapy and the antiangiogenic drugs were slowing progression.

Extensive analysis of subgroups by type of chemotherapy, performance status, sites of metastases, age set on KRAS metastasis showed no evidence of interactions [41]. Adjuvant treatment was investigated, considering the beneficial effects in advanced disease [8]. However, in five randomize trials of nearly 10,000 patients, there was no significant benefit.

This recapitulates results in other adjuvant trials and suggests that our paradigm of VEGF-induced angiogenesis is important for micrometastases and awakening from dormancy, and early growth of metastases in incorrect.

Clinical studies to investigate mechanisms of antiangiogenic response

Many studies had conducted serial investigations using imaging and molecular methods to elucidate adaptation, resistance, and response. These are vastly outnumbered by studies that have investigated plasma serum markers or polymorphisms in the VEGF signaling pathway [42]. VEGF, its soluble receptors and most of the angiogenic factors such as hepatocyte growth factor (HGF), platelet derived growth factor (PDGF), and fibroblast growth factor (FGF) members, was found to be of minimal value and none came into routine practice [43]. Gene expression profiles for hypoxia, which reflect the complex interactions between the pathways, have been investigated in radiotherapy [44] and are induced by antiangiogenic therapy. They have not been tested prospectively [45].

Baseline and serial imaging

This has been more promising and methods include ultrasound, MRI, SPECT [46], and PET labeled isotopes [47–52]. The most cost effective is ultrasound, and although operator dependent to some extent, suitable training can overcome this.

Side effects of antiangiogenic therapy

These have been extensively reviewed [53–55] for different drugs and are mainly related to specific inhibition of the drug targets, that is, not off-target effects, but unfortunately targeting normal tissue roles of VEGF and other tyrosine kinases. The normal physiology has been revealed in cases where it was not so well recognized.

A detailed review of 10,217 patients in trials of bevacizumab from 16 randomized trials was analyzed for excess fatal adverse events. Overall, there was an incidence of 2.9% in bevacizumab-treated patients and 2.2% in controls. This differed markedly by tumor type, highest in NSCLC and prostate, and no cases in a breast cancer trial [53,55]. The most common causes of fatal adverse events were hemorrhage (23.5%), neutropenia (12.2%), and gastrointestinal tract perforation (7.1%). Pulmonary embolism and gastrointestinal hemorrhage accounted for most fatal bleeding episodes. The pulmonary hemorrhaging is possibly biased because squamous cancers were included, now excluded because of that risk. It also appeared that toxicities were associated more with taxanes and platinum, but this could also reflect their use of tumor types more likely to have organ-specific toxicities such as lung and ovarian cancer. Overall, however, this is a low additional burden of 0.7%.

There are many more common toxicities to be managed, and although there is some emphasis here on bevacizumab, it should be noted that these are generic. Several reviews summarize these side effects and management [28,56], and the most important side effects will be discussed as follows.

Hypertension

This is a common side effect, routinely monitored and treated before each cycle of therapy. All the standard antihypertensives are effective, for example, calcium channel blocker, ACE inhibitors, and others. The approach is individual depending on other risk factors and drug therapy. The mechanism may be related to effects on vessel tone. An associated decline in cardiac function has been noted.

Proteinuria

This is the second most common side effect routinely assessed but no specific therapy is available. Drug is discontinued if proteinuria rises to high levels. The effects may be because of the key role of VEGF in maintaining podocyte function [57] and other effects similar to preeclampsia [58] and thrombotic microangiopathy.

Hemorrhage

This is one of the major adverse events and affects several organs. Common or minor hemorrhage occurs as hemoptysis or epistaxis, more severe may involve the lung and CNS [59]. Guidelines to prevent lung hemorrhage include to avoid treating squamous cell carcinoma and those with evidence of vessel invasion [60]. Obviously glioblastoma is of concern, and they are already associated with spontaneous hemorrhage, but severe hemorrhages occurred in only 1.3% versus 0.3% and 2% versus 0.9% of bevacizumab-treated patients versus controls in randomized trials [56]. Because of the risk of thromboembolism, some glioblastoma patients may need anticoagulants. Both heparins and warfarin have been used, and although there is an increase in incidence of CNS bleeds in these patients, it can be used [61]. For other tumor types, a metaanalysis of 8443 patients showed no increased incidence of CNS hemorrhage with bevacizumab and the conclusion was that such patients should not be excluded from trials [62].

Arterial thrombotic events

Analysis of five randomized trials [63], a twofold increase of risk with bevacizumab added to chemotherapy versus chemotherapy alone. This was only if arterial not venous thrombosis. The basis for this is likely to be similar to that for hemorrhage, with the reduction of VEGF essential for many aspects of endothelial function and integrity. The loss of endothelial covering of the subendothelial surface will expose many procoagulant signals including tissue factor and also stimulation of platelet aggregation. Arterial thromboses are a strong indication of permanent cessation of anti-VEGF therapy and appropriate management of the organ vessel involved.

Anticoagulation on anti-vascular endothelial growth factor therapy

Although there is no increase in venous thromboembolism on anti-VEGF therapy, many patients will be on chemotherapy and also will have cancer risk factors for venous thromboembolism (VTE). The preferred anticoagulants are low-molecular-weight heparins, also oral antiplatelet drugs. In a larger analysis [64] of 7956 patients, the incidence of VTE was 11.9% with a significantly higher frequency than controls. The highest risks were in colorectal and NSCLC patients, the lowest in breast and renal cancer. It is noteworthy that only three deaths from VTE were found in these trials.

Thus the need for anticoagulation will be common and it appears effective. Anticoagulation is not a contraindication to start antiangiogenic therapy and should be used as clinically indicated.

Gut perforation, fistulae, and wound healing

Gastrointestinal perforation or fistulae have serious implications of antiangiogenic therapy [65–68]. They are all probably related to the role of angiogenesis again in the wound healing process. Several risk factors are identified, clearly intraabdominal illness from colorectal and ovarian cancer, use of steroids, bowel surgery, and abdominal irradiation. Preexisting evidence of bowel obstruction, bowel wall infiltration, and inflammation such as colitis and diverticulitis are risk factors. It is generally recommended to stop anti-VEGF therapy well before surgery in accordance with time for the drug to clear. In the case of bevacizumab, this would be 6 weeks before surgery and for all drugs to restart after 4 weeks.

However, minor operations such as removal of drug administration ports were minimally affected in patients on bevacizumab. Although complication rate was higher if operation done within 1 day of a bevacizumab dose (usually given once every 3 weeks), this only amounted to 2.4% versus 0.3% when done 2–3 days later. Within 7 days it was. However, no increase was found at 14 days [69]. For bowel surgery on bevacizumab, there was a complication in wound healing in 13% versus 3.4%. Starting bevacizumab 28 days after surgery had no effects on wound healing [68].

Other side effects

Thyroid dysfunction

Hypothyroidism and subclinical [70] abnormalities are common, function should be checked regularly and hypothyroidism may explain some cases of extreme fatigue. The latter is another rarer side effect [71,72].

Pancreatitis, muscle wasting, nasal septum perforation are other side effects possibly related to normal roles of VEGF [73–75].

Abnormal liver function tests are common and usually do not require any specific intervention [76].

Cutaneous toxicity

Although rare with bevacizumab, it is a common side effect with TKIs, manifesting both as the hand–foot syndrome and as a generalized rash [77]. There is some specificity in that sorafenib was associated with hand–foot syndrome, whereas sunitinib with generalized rash. This most likely reflects their inherent TK inhibition profiles.

Posterior reversible encephalopathy syndrome, as its name suggests, is treatment with immediate cessation of anti-VEGF therapy and control of any associated hypertension. Headaches, seizures, nausea, and vomiting occur within 2–3 weeks of starting therapy and imaging shows focal vasogenic edema [78].

Multiple mechanisms of vascularization and resistance to antiangiogenic therapy

It has become increasingly clear over the last decade that "angiogenesis" is not the only way tumors develop a vascular system, and it may not even be the most common. Resistance may occur primarily because the specific molecule being targeted is not the main mechanism of angiogenesis, for example, VEGF, and attempts to overcome this have been via multitargeted kinases [79] (Fig. 9.2).

A plethora of single antibodies have been used, none approaching the single agent activity of bevacizumab. However, other vascular

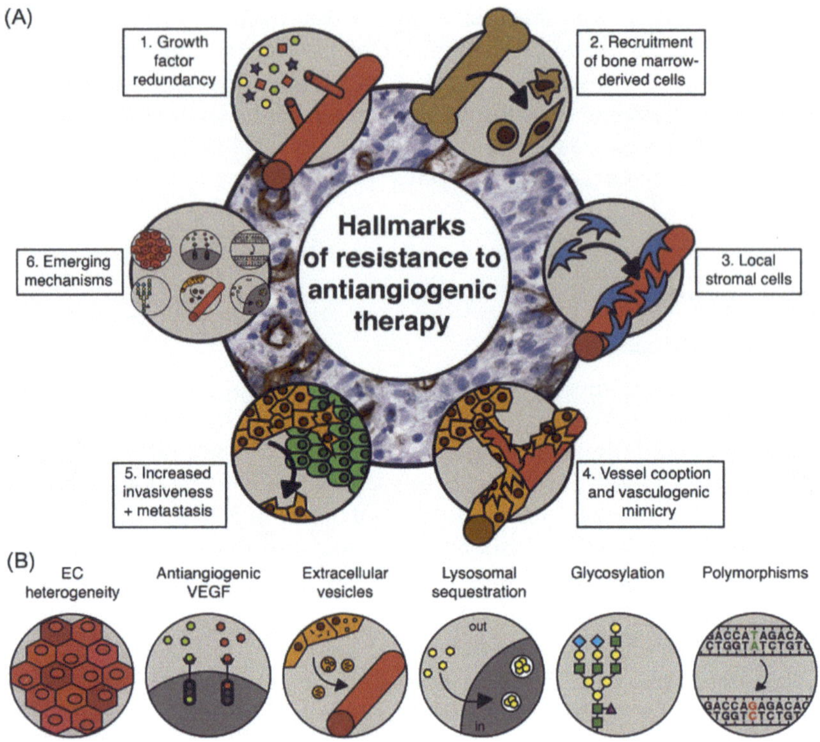

FIGURE 9.2 Multiple mechanisms of resistance to antiangiogenic drugs.
The hallmarks of resistance to antiangiogenic treatment. (A) Five distinct mechanisms to overcome antiangiogenic treatment can be distinguished. The sixth group (B) comprises a growing number of emerging mechanisms contributing to loss of activity of antiangiogenic drugs. Source: *Reproduced with permission from van Beijnum JR, Nowak-Sliwinska P, Huijbers EJ, Thijssen VL, Griffioen AW. The great escape; the hallmarks of resistance to antiangiogenic therapy. Pharmacol Rev 2015;67(2):441–61.*

targets, not necessarily angiogenic, are encouraging [80]. There are many other angiogenic factors that can cause angiogenesis and currently we have no way of defining which ones are contributing. It seems unlikely from data so far that any single one will match the role of VEGF, but some of the most obvious such, as members of the FGF family, remain to be adequately inhibited. Furthermore, the hypoxia that is often induced by antiangiogenic therapy can induce many other angiogenic factors as well as VEGF itself.

Many anticancer effects of the immune system are suppressed by hypoxia, a potential mechanism for the short duration of antitumor effects of therapy [81–85]. The immunological mechanisms of escape due to hypoxia and VEGF are being investigated as potential new therapy approaches. Other stromal cells such as fibroblasts and pericytes have key roles and their functions differ not only in different cancers but also at different sites. The hypoxia induced by therapy induces many mechanisms involved in invasion and metastasis, although there has been no convincing evidence of enhanced tumor growth in patients caused by therapy. There is one circumstance where antiangiogenic therapy is given on an intermittent basis because of toxicity, with the drug sorafenib. Here it was found that having a 2-week gap rather than a 1-week gap was, not surprisingly, detrimental. There was no indication of a more rapid regrowth than before treatment.

Tumor endothelial heterogeneity, by type, vascular bed and changed by therapy is a rapidly developing area informed by single cell sequencing. It is likely this will greatly enhance our understanding and therapeutics in the near future. As yet, none of these is routinely considered in patient selection or for second-line therapy, but as the new mechanisms are understood and imaging modalities are at higher resolution, it is expected that refined clinical management will be possible.

Metabolic adaptation to antiangiogenic drugs and role of hypoxia-inducible factor

Hypoxia is a key physiological difference between tumor and normal tissue and is at the interface of tumor-induced angiogenesis and effectiveness and of antiangiogenic therapy. The role and mechanisms of activation of the HIF1 and HIF2 have been extensively reviewed and will not be further reviewed here [86,87]. Thus the hypoxia generated by tumor proliferation and oxygen consumption [88], coupled with poorly functioning vessels generated in response to VEGF and other factors have many protumorigenic effects.

The induction of HIF regulates many key pathways of metabolism which enhance survival [89] in Fig. 9.3.

Key is glycolysis, but also fatty acid uptake and increase in antioxidants (Fig. 9.4).

Lactate is a major metabolite whose production is induced by hypoxia. It is transported via lactate transporters MC74 and MCT1. Additionally, $H+$ ion excretion is facilitated by carbonic anhydrase 9. Inhibitors of both types of transporter and enzyme are in phase I trials. Inhibitors of CA9 greatly potentiated the effects of antiangiogenic therapy in preclinical studies [90]. It is likely that the adaptation to antiangiogenic drugs will vary and from patient to patient, so imagining of hypoxia and angiogenesis in clinical studies is important to improve therapy.

Recent approaches in breast cancer typify how to investigate the biology in patients. In a study of 70 TRN patients, basal MV, gene expression, interstitial fluid pressure, HIF1α expression, Ki67, and αSMA coated

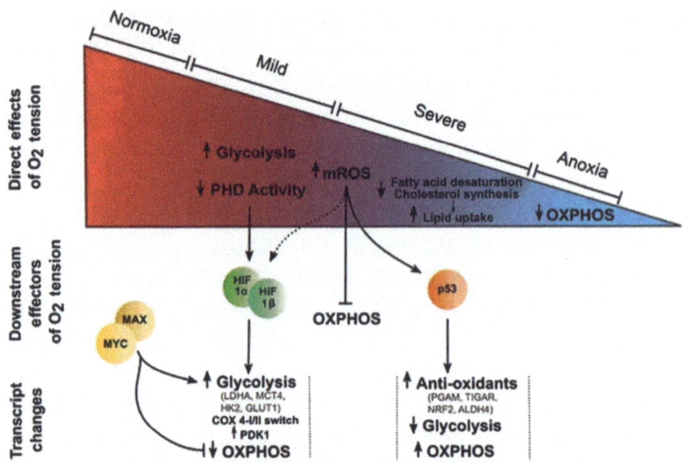

FIGURE 9.3 **Metabolic effects of hypoxia.** Decreasing oxygen tension elicits alterations in metabolism through a number of mechanisms. The direct effects of oxygen tension on metabolism are shown within the "oxygenation wedge," whereas those that are affected through signaling pathways activated in hypoxia are shown below. *ALDH4*, Aldehyde dehydrogenase 4; *GLUT1*, facultative glucose transporter 1; *HIF*, hypoxia-inducible factor; *HK2*, hexokinase 2; *LDHA*, lactate dehydrogenase A; *MAX*, Myc-associated protein X; *MCT4*, monocarboxylate transporter 4; *mROS*, mitochondrial reactive oxygen species; *MYC*, V-Myc Avian Myelocytomatosis Viral Oncogene Homologue; *NRF2*, nuclear factor erythroid 2-related factor 2; *OXPHOS*, oxidative phosphorylation; *PDK1*, pyruvate dehydrogenase kinase 1; *PGAM*, phosphoglycerate mutase. Source: *From Eales KL, Hollinshead KER, Tennant DA. Hypoxia and metabolic adaptation of cancer cells. Oncogenesis 2016;5:190. doi:10.1038/oncsis.2015.50. https://creativecommons.org/licenses/by/4.0/.*

vessels were measured [91,92]. These gave indications of hypoxia proliferation and vascular normalization (pericyte coverage). Patients received chemotherapy with adriamycin and cyclophosphamide for 4 × 3-week cycles, then taxol for 4 × 3-week cycles, followed by surgery. In the 2-week window, the studies were done with the bevacizumab (BEV) alone, which continued to be administered throughout chemotherapy. Microvessel density (MVD) before treatment gave the best correlation with final pathological response. Similarly, the increase in pericyte-covered vessels after BEV correlated with pathological CR. The decrease in IFF that occurred was less than reported previously in rectal cancer. This generated the hypothesis that BEV was only likely to be effective in those with high MVD, which may then be reduced and normalized with beneficial effects for drug delivery. Those with low MVD may have even less vessels and become more hypoxic.

Harris's group did a similar window study prior to neoadjuvant chemotherapy, but BEV was not continued after 1 cycle [93]. In that study,

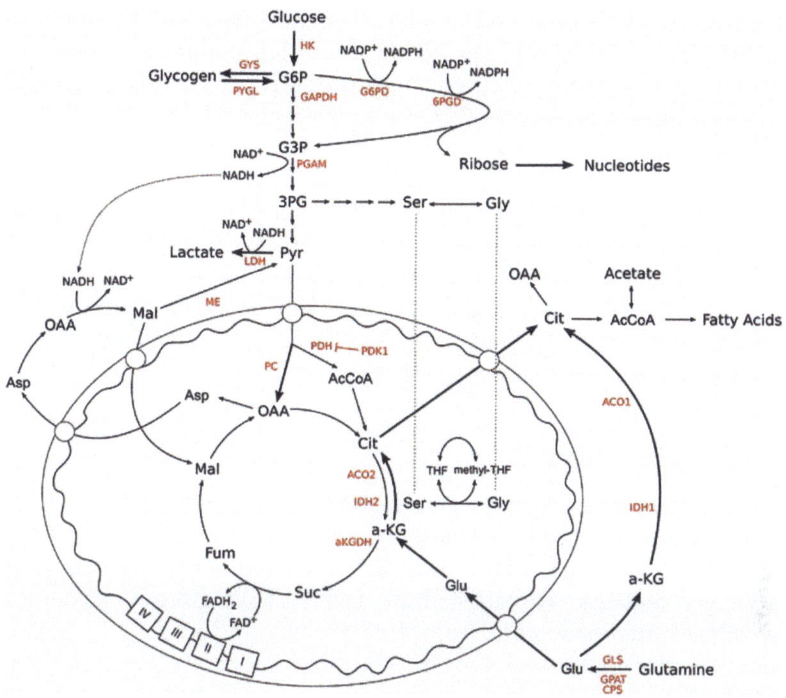

FIGURE 9.4 **Metabolic pathways modulated in hypoxia.** Some of the metabolic pathways known to have altered flux or importance in hypoxia. Key enzymes metabolic pathways described are shown in red. Thicker lines represent those pathways in which hypoxic cells have been shown to increase flux, or rely more on their activity. *αKG*, Alpha-ketoglutarate; *3PG*, 3-phosphoglycerate; *6PGD*, 6-phosphogluconate dehydrogenase; *AcCoA*, Acetyl-Coenzyme A; *ACO1/2*, aconitase 1/2; *ALD*, aldolase; *Asp*, aspartate; *Cit*, citrate; *CPS*, cytidine triphosphate synthetase; *Fum*, fumarate; *G3P*, glyceraldehyde 3-phosphate; *G6P*, glucose 6-phosphate; *G6PD*, glucose 6-phosphate dehydrogenase; *GAPDH*, glyceraldehyde 3-phosphate dehydrogenase; *Glu*, glutamate; *GLS*, glutaminase; *Gly*, glycine; *GPAT*, glutamine phosphoribosylpyrophosphate amidotransferase; *GYS*, glycogen synthase; *HK*, hexokinase; *IDH1/2*, isocitrate dehydrogenase 1/2; *LDH*, lactate dehydrogenase; *Mal*, malate; *ME*, malic enzyme; *OAA*, oxaloacetate; *PC*, pyruvate carboxylase; *PDH*, pyruvate dehydrogenase; *PDK1*, pyruvate dehydrogenase kinase 1; *PGAM*, phosphoglycerate mutase; *PYGL*, glycogen phosphorylase; *Pyr*, pyruvate; *Ser*, serine; *Suc*, succinate. Source: *From Eales KL, Hollinshead KER, Tennant DA. Hypoxia and metabolic adaptation of cancer cells. Oncogenesis 2016;5:190. doi:10.1038/oncsis.2015.50. https://creativecommons.org/licenses/by/4.0/.*

there was a clear reduction in Ktrans measured by MRI and those with highest levels showed the greatest reduction. There was a highly significant relationship between the fall in Ktrans and induction of HIF-dependent genes measured by exon arrays. MVD fell, as did Ki67, with an increase in HIF1α and CAIX. This showed that the effect of BEV was to produce hypoxia and adaptation within 2 weeks. Amongst the genes induced were those involved in glycolysis and many angiogenic genes, including the target VEGF. The reduction of Ktrans showed that permeability was decreased, so evidence of "normalization," but also hypoxia increased. This study emphasizes the rapid feedback to overcome angiogenesis inhibition and suggests that inhibition of several of the induced pathways should be used in combination. Additionally, many genes involved in immunosuppression were increased, although VEGF has many immunosuppressive effects. It is likely this reflects effects of hypoxia and supports the use of combination ICB with antiangiogenic drugs. Other studies have used this approach, but assessment was done in combination with chemotherapy, making the independent role of bevacizumab difficult to assess [43].

Induced essentiality

The adaptation to hypoxia and metabolic changes such as increased glycolysis, reduced lipid and protein synthesis, adaptation to the acid pH induced by lactate [94], and carbonic acid and monocarboxylate transporters required to redistribute lactate have been well reviewed. Many other transporters and receptors are also induced by hypoxia and present essential pathways for survival, and hence are potential targets. Amongst these, the enzyme carbonic anhydrase IX has been extensively studied, not only as a marker of hypoxia and HIF signaling, but as a therapy target. A compound inhibiting this specific isoform, U104, is currently in clinical trials. Other targets include LOXL2, CXCR4, MCT, and an MCT1 inhibitor are being assesses with pharmacodynamic endpoints.

Stem cells and hypoxia

Stem cells are routinely grown in mild hypoxia as this enhances their growth. Many of the canonical stem cell transcription factors are induced by hypoxia via HIF1alpha or HIF2alpha [95–99]. A concerning aspect of antiangiogenic therapy is the potential for selecting stem cells under hypoxia [95,100,101]. Substantial evidence [102,103] in stem cell niches in the bone marrow, oxygen tension is below 1%, under physiological hypoxia in tumors is around 2%–9%. In the early embryo, oxygen gradients are essential for development, which occurs in a relatively oxygen poor environment. Hypoxia slows down proliferation and helps maintain them in an undifferentiated state. This appears to be the case for normal tissue and also for cancer stem cells. HIF2 is more important than HIF1 in several studies regulating Oct4, SOX2, and NANOG, and inhibition of HIF2 prevents in vivo growth. Notch signaling has a key role, regulates EMT, and is required for hypoxia to maintain an undifferentiated state [102,103]. HIF1 also has a role via notch in maintaining LSC self-renewal in lymphoma and acute myeloid leukemia [104]. Most work has been done on long-term maintenance of the leukemic cells [105]. In cancer cells, lines HIF2 induce stem cell phenotypes by activating Wnt and Notch [101]. Other mechanisms include induction of cytokines activating STAT3, and ALDH1A1 inducing VEGF production [106]. In breast cancer, hypoxia activated NANOG through ALHBJ-mediated demethylation of NANONG mRNA. Thus combination of drugs targeting stem cell properties, such as Wnt and Notch, will be of interest.

New approaches to combination therapy

Hypoxia activated prodrugs

This class of drug has been investigated in clinical trials for nearly 25 years. They are activated by cellular oxido-reductases to toxic metabolites, which is a reversible reaction in the presence of oxygen [107,108]. Three main drugs have been tested in randomized trials, all considered to have failed. The first was tirapazamine (SR4233) [109] in head and neck, lung cancer, and cervical cancer. The drug failed to show benefit when added to radiotherapy or cisplatin or both. The reasons included toxicity and need for dose adjustment, inadequate standardization of radiotherapy, unusual side effects, such as sudden deafness, and new combination of chemotherapy superseding any benefit of the combination.

PR-104 was a later drug, tested in the clinic but caused marrow toxicity and was found to be activated by an aldo-keto reductase, which can reduce PR-104 to its active form regardless of oxygen [110].

TH-302 (evofosfamide) [110] seemed to have overcome many of the previous flaws and was analyzed in two randomized trials, one on soft-tissue sarcoma with doxorubicin (SARCO21), the other in advanced pancreatic cancer with gemcitabine (MAESTRO). Neither showed a significant effect.

The problem with all these trials is the failure to stratify patients by the degree of hypoxia in their tumors, which varies greatly. Several groups have described biomarkers such as carbonic anhydrase IX and gene expression profiles [44] and imaging by PET scans or mother modalities using misonidazole (FMISO), FAZA or HX4 or MRI [47–49,51,52,111,112].

However, several new drugs are being designed, such as hypoxia-activated DNA repair inhibitors or antiangiogenic drugs, which could focus toxicity to the tumor and synergize in a different way to the cytotoxic drugs already used [113].

Additionally, clinical trials of hypoxia-activated pan-ErbB inhibitor tarloxotinib will use PET imaging of hypoxia [108]. Thus the combination with antiangiogenics will be tested in the near future with appropriate biomarkers and sequencing design.

DNA repair inhibitors are of particular interest because of the evidence for accumulation of DNA single strands in hypoxia and effects of reoxygenation of DNA damage induction [114,115] and synergy with radiation.

Selective antibody activation proantibodies

Considering the importance of antibodies to antiangiogenic therapy, new approaches to activate them selectively in the tumor microenvironment offer considerable potential. *Probodys* are based on the principle of blocking the antigen binding sites of the antibody with a peptide that is cleavable in the correct microenvironment by specific proteases. A recent approach with anti-EGFR antibodies and anti-TNFα used a human protein inhibitor domain of latency-associated peptide, domains C2b or CBa of complement factor 2/B, linked by a peptide that could be cleaved by MMP2 [116]. Results showed much greater selectivity and reduced toxicity. Many proteases are induced in hypoxia and acidic environments, which would be induced by antiangiogenic therapy.

Drug delivery

Although beyond the scope of this review, advances in nanotechnology are being exploited specifically for antiangiogenic therapy, with advantages for the great variety of payloads that can be delivered, as well as tissue [117]. Particles range from 1 nm upwards and extravasate through the leaky tumor vasculature, although vascular cooption and other mechanisms of vascularization may not exhibit this effect. The enhanced permeability and retention effect may also be complemented by specific targeting modules.

Antiangiogenesis to overcome resistance to immunotherapy

The largest change in cancer therapy in the last 5 years has been the success of immunotherapy with ICB. As an example, pembrolizumab (Keytruda), an antiprogrammed cell death 1 receptor antibody, is now licensed for the following cancer types: melanoma, NSCLC, head and neck squamous cancer, Hodgkin lymphoma, primary mediastinal large B-cell lymphoma, urothelial carcinoma, microsatellite instability-high cancer, gastric cancer, cervical cancer, hepatocellular cancer, Merkel cell cancer, and renal cell carcinoma. Similar indications are found for many other inhibitors of this class. In many cases, this has become first- or second-line treatment, in the same tumor types for which antiangiogenics were also indicated as first or second line. Clearly, these immunomodulators are far more effective, yet the majority of patients will not benefit from them. Hence understanding resistance and combination therapy is critical to improve outcome.

There are strong mechanistic interactions between the tumor endothelium, access of immune cells to the tumor and also immune-suppressive effects of angiogenic factors, particularly established for VEGF (see Section Renal cancer). VEGF suppresses dendritic cell differentiation, function and can increase PDL1 expression [118] (Fig. 9.5).

The migration of myeloid suppressor cells is enhanced as well as Treg cell function and proliferation. Conversely, T-cell effector function and proliferation is decreased. There are many other angiogenic factors that can contribute to the suppression, for example, HGF and angiopoietins.

Additionally, the cell surface expression of cell adhesion molecules on tumor endothelium such as ICAM-1 can increase leukocyte infiltration, although in many cases ICAM-2 is downregulated and would reduce infiltration. There is great heterogeneity within tumors and between them in the expression of multiple cell adhesion molecules, likely to modulate the tumor infiltrate. Immunosuppressive molecules such as PDL1 are upregulated on human tumor endothelial cells [119,120]. The overall result is that the tumor endothelium can produce an immunological barrier and "cold" tumors. As a consequence of this interaction, dozens of studies are now ongoing using different immune checkpoint inhibitors in combination with antiangiogenic drugs in the tumor types for which they are already licensed individually. An example is [121] esophageal-gastric cancer, where ramucirumab is licensed in combination with anti-PDL1 therapy.

One of the mechanisms, by which antiangiogenic therapy may enhance immune response, is vascular normalization, which could reduce tumor hypoxia (itself a major immune suppressor) and enhance T-cell infiltration. Decreased Tregs, increased polarization of macrophages to the MI phenotype and downregulation of PDL1 (an HIF target), have been demonstrated [122]. Tian et al. [123] showed that there was a surprising reciprocal regulation of vascular normalization and immune stimulatory reprogramming. Activation of CD4 + T-cells, particularly of TH1 cells, increased vessel normalization through IFγ secretion and increasing

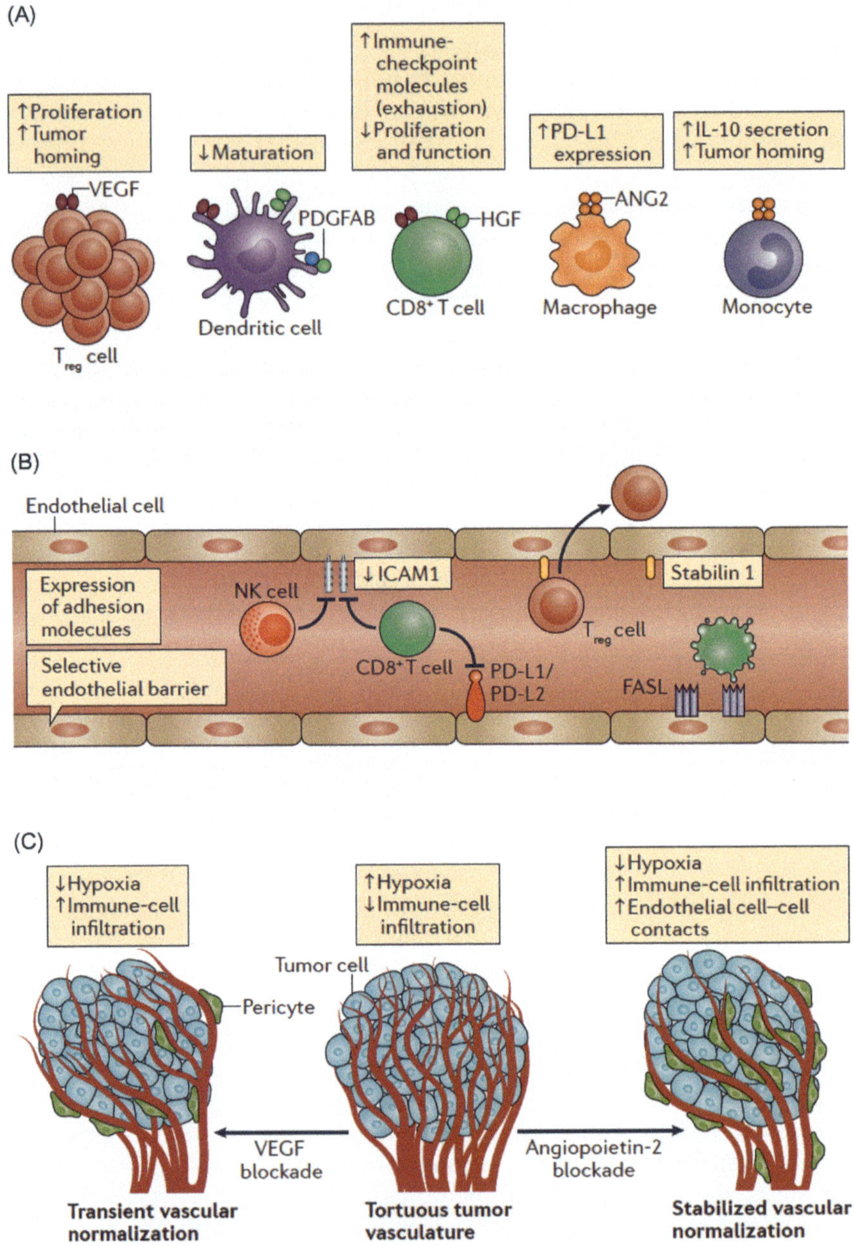

FIGURE 9.5 **Effects of antiangiogenic drugs and hypoxia on the immune system.** Angiogenesis-modulating factors have effects on the immune system in three established ways. (A) VEGF can increase both regulatory T (Treg) cell proliferation and homing to tumor tissues. VEGF can also suppress dendritic cell maturation and CD8 + T-cell proliferation and function and cause T-cell exhaustion. Angiopoietin 2 (ANG2) can bind to

pericyte coverage of vessels. IFNγ secretion is a critical feature of ICB therapies.

Disruption of vessel normalization reduces T-lymphocyte infiltration. Thus vascular normalization by antiangiogenic therapy could enhance T-cell entry which would reinforce the normalization and immune response, and vice versa [124]. Dual blockade in hepatoma models of PD-1 and VEGFR2 did indeed show a strong synergistic effect via vascular normalization, an approach now in a clinical trial for hepatoma [125]. This combination of anti-PDL1 and anti-VEGFR2 therapy also enhanced formula of a specific type of vessels, high endothelial venules, which are a route for lymphocyte infiltration [126].

The induction of perfusion by ICB alone was a predictor of response which would also reduce hypoxia [123,127]. These data are helpful in attempts to classify patients into those most likely to benefit from the combinations [128] but have not been prospectively validated. The multiple additional mechanisms by which modulation of angiogenesis and hypoxia improve the immune response have been well reviewed [129,130].

The likelihood is that many more combinations will be approved, and as expense increases the importance of biomarkers will increase. This may be more to select patients for the correct combinations and to predict resistance and offer new opportunities. As the response rates go up, prediction will be less important overall, but classifications into mechanistically determined therapy would increase cost effectiveness, hopefully response and reduce side effects.

◀ macrophages and monocytes, resulting in immunosuppression. HGF, as well as PDGFAB, can bind to dendritic cells, thus suppressing their maturation. HGF can also bind to T cells and suppress effector T cell function. (B) Through the regulation of the expression of adhesion molecules, certain immunosuppressive cells can be allowed into tumor tissues (e.g., stabilin 1-mediated Treg cell trafficking) and the infiltration of certain effector cells into tumors can be blocked [e.g., intercellular adhesion molecule 1 (ICAM1) downregulation leads to the suppression of natural killer (NK) cell and T-cell trafficking]. Selective endothelial barriers can be created when the expression of molecules that either suppress effector cell function [such as the immune-checkpoint molecules programmed cell death 1 ligand 1 (PD-L1) and 2 (PD-L2)] or cause effector cell apoptosis [such as FAS antigen ligand (FASL)] is deregulated. (C) Vascular normalization can result in indirect physical effects, which lead to reduced hypoxia and increased immune-cell infiltration. Upon low-level VEGF blockade, the tortuous tumor vasculature becomes transiently normalized, with more-regular vessel patterning and pericyte coverage. ANG2 blockade results in prolonged vessel normalization, characterized by increased stability of endothelial cell–cell contacts and pericyte coverage as well as by the presence of enlarged vessels with fewer branches. Source: *Reproduced with permission from Khan KA, Kerbel RS. Improving immunotherapy outcomes with anti-angiogenic treatments and vice versa. Nat Rev Clin Oncol 2018;15 (5):310–24.*

References

[1] Jayson GC, Kerbel R, Ellis LM, Harris AL. Antiangiogenic therapy in oncology: current status and future directions. Lancet 2016;388(10043):518–29.
[2] Ye W. The complexity of translating anti-angiogenesis therapy from basic science to the clinic. Dev Cell 2016;37(2):114–25.
[3] Lupo G, et al. Anti-angiogenic therapy in cancer: downsides and new pivots for precision medicine. Front Pharmacol 2016;7:519.
[4] Klein D. The tumor vascular endothelium as decision maker in cancer therapy. Front Oncol 2018;8:367.
[5] Wang Z, et al. Broad targeting of angiogenesis for cancer prevention and therapy. Semin Cancer Biol 2015;35(Suppl):S224–43.
[6] Fox SB, et al. Relationship of endothelial cell proliferation to tumor vascularity in human breast cancer. Cancer Res 1993;53(18):4161–3.
[7] Rahaman MM, Straub AC. The emerging roles of somatic globins in cardiovascular redox biology and beyond. Redox Biol 2013;1:405–10.
[8] Kim BJ, Jeong JH, Kim JH, Kim HS, Jang HJ. The role of targeted agents in the adjuvant treatment of colon cancer: a meta-analysis of randomized phase III studies and review. Oncotarget 2017;8(19):31112–18.
[9] Jain RK. Normalization of tumor vasculature: an emerging concept in antiangiogenic therapy. Science 2005;307(5706):58–62.
[10] Martin JD, Seano G, Jain RK. Normalizing function of tumor vessels: progress, opportunities, and challenges. Annu Rev Physiol 2019;81:505–34.
[11] Jain RK. Normalizing tumor vasculature with anti-angiogenic therapy: a new paradigm for combination therapy. Nat Med 2001;7(9):987–9.
[12] Donovan P, et al. Endovascular progenitors infiltrate melanomas and differentiate towards a variety of vascular beds promoting tumor metastasis. Nat Commun 2019;10(1):18.
[13] He L, et al. Single-cell RNA sequencing of mouse brain and lung vascular and vessel-associated cell types. Sci Data 2018;5:180160.
[14] Zecchin A, Kalucka J, Dubois C, Carmeliet P. How endothelial cells adapt their metabolism to form vessels in tumors. Front Immunol 2017;8:1750.
[15] Raphael J, et al. Antiangiogenic therapy in advanced non-small-cell lung cancer: a meta-analysis of phase III randomized trials. Clin Lung Cancer 2017;18(4) 345-353.e345.
[16] Janning M, Loges S. Anti-angiogenics: their value in lung cancer therapy. Oncol Res Treat 2018;41(4):172–80.
[17] Kim M, et al. Major clinical research advances in gynecologic cancer in 2018. J Gynecol Oncol 2019;30(2):e18.
[18] Yoshida H, Yabuno A, Fujiwara K. Critical appraisal of bevacizumab in the treatment of ovarian cancer. Drug Des Devel Ther 2015;9:2351–8.
[19] Rossi L, et al. Bevacizumab in ovarian cancer: a critical review of phase III studies. Oncotarget 2017;8(7):12389–405.
[20] Miller K, et al. Paclitaxel plus bevacizumab versus paclitaxel alone for metastatic breast cancer. N Engl J Med 2007;357(26):2666–76.
[21] Wagner AD, Thomssen C, Haerting J, Unverzagt S. Vascular-endothelial-growth-factor (VEGF) targeting therapies for endocrine refractory or resistant metastatic breast cancer. Cochrane Database Syst Rev 2012;(7):CD008941.
[22] Nahleh Z, et al. Bevacizumab in the neoadjuvant treatment of human epidermal growth factor receptor 2-negative breast cancer: a meta-analysis of randomized controlled trials. Mol Clin Oncol 2019;10(3):357–65.
[23] Cao L, et al. Neoadjuvant bevacizumab plus chemotherapy versus chemotherapy alone to treat non-metastatic breast cancer: a meta-analysis of randomised controlled trials. PLoS One 2015;10(12):e0145442.

[24] Martin M, et al. Evaluating the addition of bevacizumab to endocrine therapy as first-line treatment for hormone receptor-positive metastatic breast cancer: a pooled analysis from the LEA (GEICAM/2006-11_GBG51) and CALGB 40503 (Alliance) trials. Eur J Cancer 2019;117:91–8.

[25] Schmid P, et al. Atezolizumab and Nab-Paclitaxel in advanced triple-negative breast cancer. N Engl J Med 2018;379(22):2108–21.

[26] Vinayak S, et al. Open-label clinical trial of niraparib combined with pembrolizumab for treatment of advanced or metastatic triple-negative breast cancer. JAMA Oncol 2019. Available from: https://doi.org/10.1001/jamaoncol.2019.1029.

[27] Alonso-Gordoa T, et al. Targeting tyrosine kinases in renal cell carcinoma: "new bullets against old guys". Int J Mol Sci 2019;20(8).

[28] Eisen T, et al. Targeted therapies for renal cell carcinoma: review of adverse event management strategies. J Natl Cancer Inst 2012;104(2):93–113.

[29] Jain RK, Gandhi S, George S. Second-line systemic therapy in metastatic renal-cell carcinoma: a review. Urol Oncol 2017;35(11):640–6.

[30] Tzogani K, et al. The European Medicines Agency approval of axitinib (Inlyta) for the treatment of advanced renal cell carcinoma after failure of prior treatment with sunitinib or a cytokine: summary of the scientific assessment of the committee for medicinal products for human use. Oncologist 2015;20(2):196–201.

[31] Motzer RJ, et al. Nivolumab plus ipilimumab versus sunitinib in first-line treatment for advanced renal cell carcinoma: extended follow-up of efficacy and safety results from a randomised, controlled, phase 3 trial. Lancet Oncol 2019;20(10):1370–85.

[32] Rini BI, et al. Atezolizumab plus bevacizumab versus sunitinib in patients with previously untreated metastatic renal cell carcinoma (IMmotion151): a multicentre, open-label, phase 3, randomised controlled trial. Lancet 2019;393(10189):2404–15.

[33] Ameratunga M, et al. Anti-angiogenic therapy for high-grade glioma. Cochrane Database Syst Rev 2018;11:CD008218.

[34] Elder JB, Nahed BV, Linskey ME, Olson JJ. Congress of neurological surgeons systematic review and evidence-based guidelines on the role of emerging and investigational therapties for the treatment of adults with metastatic brain tumors. Neurosurgery 2019;84(3):E201–e203.

[35] Sorensen AG, et al. A "vascular normalization index" as potential mechanistic biomarker to predict survival after a single dose of cediranib in recurrent glioblastoma patients. Cancer Res 2009;69(13):5296–300.

[36] Nia HT, Munn LL, Jain RK. Mapping physical tumor microenvironment and drug delivery. Clin Cancer Res 2019;25(7):2024–6.

[37] Ly KI, et al. Probing tumor microenvironment in patients with newly diagnosed glioblastoma during chemoradiation and adjuvant temozolomide with functional MRI. Sci Rep 2018;8(1):17062.

[38] Batchelor TT, et al. Improved tumor oxygenation and survival in glioblastoma patients who show increased blood perfusion after cediranib and chemoradiation. Proc Natl Acad Sci U S A 2013;110(47):19059–64.

[39] Bonekamp D, et al. Assessment of tumor oxygenation and its impact on treatment response in bevacizumab-treated recurrent glioblastoma. J Cereb Blood Flow Metab 2017;37(2):485–94.

[40] Wagner AD, Arnold D, Grothey AA, Haerting J, Unverzagt S. Anti-angiogenic therapies for metastatic colorectal cancer. Cochrane Database Syst Rev 2009;(3):Cd005392.

[41] Hurwitz HI, et al. Efficacy and safety of bevacizumab in metastatic colorectal cancer: pooled analysis from seven randomized controlled trials. Oncologist 2013;18(9):1004–12.

[42] de Haas S, et al. Genetic variability of VEGF pathway genes in six randomized phase III trials assessing the addition of bevacizumab to standard therapy. Angiogenesis 2014;17(4):909–20.

[43] Kimbung S, et al. Assessment of early response biomarkers in relation to long-term survival in patients with HER2-negative breast cancer receiving neoadjuvant chemotherapy plus bevacizumab: results from the Phase II PROMIX trial. Int J Cancer 2018;142(3):618–28.
[44] Toustrup K, et al. Validation of a 15-gene hypoxia classifier in head and neck cancer for prospective use in clinical trials. Acta Oncol 2016;55(9–10):1091–8.
[45] Eustace A, et al. A 26-gene hypoxia signature predicts benefit from hypoxia-modifying therapy in laryngeal cancer but not bladder cancer. Clin Cancer Res 2013;19(17):4879–88.
[46] Chen B, et al. (99m)Tc-3P-RGD2 SPECT to monitor early response to bevacizumab therapy in patients with advanced non-small cell lung cancer. Int J Clin Exp Pathol 2015;8(12):16064–72.
[47] Ueda S, et al. In vivo imaging of eribulin-induced reoxygenation in advanced breast cancer patients: a comparison to bevacizumab. Br J Cancer 2016;114(11):1212–18.
[48] Patel N, Harris AL, Gleeson FV, Vallis KA. Clinical imaging of tumor angiogenesis. Future Oncol 2012;8(11):1443–59.
[49] Fournier L, et al. Imaging response of antiangiogenic and immune-oncology drugs in metastatic renal cell carcinoma (mRCC): current status and future challenges. Kidney Cancer 2017;1(2):107–14.
[50] Dubois LJ, et al. New ways to image and target tumour hypoxia and its molecular responses. Radiother Oncol 2015;116(3):352–7.
[51] Chen BB, et al. Imaging biomarkers from multiparametric magnetic resonance imaging are associated with survival outcomes in patients with brain metastases from breast cancer. Eur Radiol 2018;28(11):4860–70.
[52] Chen H, Niu G, Wu H, Chen X. Clinical application of radiolabeled RGD peptides for PET imaging of integrin alphavbeta3. Theranostics 2016;6(1):78–92.
[53] Ranpura V, Hapani S, Wu S. Treatment-related mortality with bevacizumab in cancer patients: a meta-analysis. JAMA 2011;305(5):487–94.
[54] Arnold D, et al. Meta-analysis of individual patient safety data from six randomized, placebo-controlled trials with the antiangiogenic VEGFR2-binding monoclonal antibody ramucirumab. Ann Oncol 2017;28(12):2932–42.
[55] Schutz FA, Je Y, Richards CJ, Choueiri TK. Meta-analysis of randomized controlled trials for the incidence and risk of treatment-related mortality in patients with cancer treated with vascular endothelial growth factor tyrosine kinase inhibitors. J Clin Oncol 2012;30(8):871–7.
[56] Brandes AA, Bartolotti M, Tosoni A, Poggi R, Franceschi E. Practical management of bevacizumab-related toxicities in glioblastoma. Oncologist 2015;20(2):166–75.
[57] Izzedine H, et al. VEGF signalling inhibition-induced proteinuria: mechanisms, significance and management. Eur J Cancer 2010;46(2):439–48.
[58] Patel TV, et al. A preeclampsia-like syndrome characterized by reversible hypertension and proteinuria induced by the multitargeted kinase inhibitors sunitinib and sorafenib. J Natl Cancer Inst 2008;100(4):282–4.
[59] Qi WX, et al. Incidence and risk of hemorrhagic events with vascular endothelial growth factor receptor tyrosine-kinase inhibitors: an up-to-date meta-analysis of 27 randomized controlled trials. Ann Oncol 2013;24(12):2943–52.
[60] Reck M, et al. Predicting and managing the risk of pulmonary haemorrhage in patients with NSCLC treated with bevacizumab: a consensus report from a panel of experts. Ann Oncol 2012;23(5):1111–20.
[61] Norden AD, et al. Safety of concurrent bevacizumab therapy and anticoagulation in glioma patients. J Neurooncol 2012;106(1):121–5.
[62] Besse B, et al. Bevacizumab safety in patients with central nervous system metastases. Clin Cancer Res 2010;16(1):269–78.

[63] Scappaticci FA, et al. Arterial thromboembolic events in patients with metastatic carcinoma treated with chemotherapy and bevacizumab. J Natl Cancer Inst 2007;99 (16):1232–9.
[64] Nalluri SR, Chu D, Keresztes R, Zhu X, Wu S. Risk of venous thromboembolism with the angiogenesis inhibitor bevacizumab in cancer patients: a meta-analysis. JAMA 2008;300(19):2277–85.
[65] Ganapathi AM, Westmoreland T, Tyler D, Mantyh CR. Bevacizumab-associated fistula formation in postoperative colorectal cancer patients. J Am Coll Surg 2012;214 (4):582–8 discussion 588–590.
[66] Hapani S, Chu D, Wu S. Risk of gastrointestinal perforation in patients with cancer treated with bevacizumab: a meta-analysis. Lancet Oncol 2009;10(6):559–68.
[67] Qi WX, Sun YJ, Tang LN, Shen Z, Yao Y. Risk of gastrointestinal perforation in cancer patients treated with vascular endothelial growth factor receptor tyrosine kinase inhibitors: a systematic review and meta-analysis. Crit Rev Oncol Hematol 2014;89 (3):394–403.
[68] Scappaticci FA, et al. Surgical wound healing complications in metastatic colorectal cancer patients treated with bevacizumab. J Surg Oncol 2005;91(3):173–80.
[69] Erinjeri JP, et al. Timing of administration of bevacizumab chemotherapy affects wound healing after chest wall port placement. Cancer 2011;117(6):1296–301.
[70] Rini BI, et al. Hypothyroidism in patients with metastatic renal cell carcinoma treated with sunitinib. J Natl Cancer Inst 2007;99(1):81–3.
[71] Fujiwara Y, et al. Management of axitinib (AG-013736)-induced fatigue and thyroid dysfunction, and predictive biomarkers of axitinib exposure: results from phase I studies in Japanese patients. Invest New Drugs 2012;30(3):1055–64.
[72] Ghatalia P, et al. Fatigue with vascular endothelial growth factor receptor tyrosine kinase inhibitors and mammalian target of rapamycin inhibitors in patients with renal cell carcinoma (RCC) and other malignancies: a meta-analysis of randomized clinical trials. Crit Rev Oncol Hematol 2015;95(2):251–63.
[73] Antoun S, et al. Association of skeletal muscle wasting with treatment with sorafenib in patients with advanced renal cell carcinoma: results from a placebo-controlled study. J Clin Oncol 2010;28(6):1054–60.
[74] Ghatalia P, et al. Pancreatitis with vascular endothelial growth factor receptor tyrosine kinase inhibitors. Crit Rev Oncol Hematol 2015;94(1):136–45.
[75] Mailliez A, et al. Nasal septum perforation: a side effect of bevacizumab chemotherapy in breast cancer patients. Br J Cancer 2010;103(6):772–5.
[76] Ghatalia P, et al. Hepatotoxicity with vascular endothelial growth factor receptor tyrosine kinase inhibitors: a meta-analysis of randomized clinical trials. Crit Rev Oncol Hematol 2015;93(3):257–76.
[77] Massey PR, Okman JS, Wilkerson J, Cowen EW. Tyrosine kinase inhibitors directed against the vascular endothelial growth factor receptor (VEGFR) have distinct cutaneous toxicity profiles: a meta-analysis and review of the literature. Support Care Cancer 2015;23(6):1827–35.
[78] Seet RC, Rabinstein AA. Clinical features and outcomes of posterior reversible encephalopathy syndrome following bevacizumab treatment. QJM 2012;105(1):69–75.
[79] van Beijnum JR, Nowak-Sliwinska P, Huijbers EJ, Thijssen VL, Griffioen AW. The great escape; the hallmarks of resistance to antiangiogenic therapy. Pharmacol Rev 2015;67(2):441–61.
[80] de Bono JS, et al. Tisotumab vedotin in patients with advanced or metastatic solid tumours (InnovaTV 201): a first-in-human, multicentre, phase 1-2 trial. Lancet Oncol 2019;20(3):383–93.
[81] Hsu TS, Lai MZ. Hypoxia-inducible factor 1alpha plays a predominantly negative role in regulatory T cell functions. J Leukoc Biol 2018;104(5):911–18.

[82] Lequeux A, et al. Impact of hypoxic tumor microenvironment and tumor cell plasticity on the expression of immune checkpoints. Cancer Lett 2019;458:13–20.
[83] Li Y, Patel SP, Roszik J, Qin Y. Hypoxia-driven immunosuppressive metabolites in the tumor microenvironment: new approaches for combinational immunotherapy. Front Immunol 2018;9:1591.
[84] Vaupel P, Multhoff G. Hypoxia-/HIF-1alpha-driven factors of the tumor microenvironment impeding antitumor immune responses and promoting malignant progression. Adv Exp Med Biol 2018;1072:171–5.
[85] Wang YA, et al. Effects of tumor metabolic microenvironment on regulatory T cells. Mol Cancer 2018;17(1):168.
[86] Semenza GL. HIF-1 mediates metabolic responses to intratumoral hypoxia and oncogenic mutations. J Clin Invest 2013;123(9):3664–71.
[87] Choudhry H, Harris AL. Advances in hypoxia-inducible factor biology. Cell Metab 2018;27(2):281–98.
[88] Ulivi P, Marisi G, Passardi A. Relationship between hypoxia and response to antiangiogenic therapy in metastatic colorectal cancer. Oncotarget 2016;7(29):46678–91.
[89] Eales KL, Hollinshead KE, Tennant DA. Hypoxia and metabolic adaptation of cancer cells. Oncogenesis 2016;5:e190.
[90] McIntyre A, Harris AL. Metabolic and hypoxic adaptation to anti-angiogenic therapy: a target for induced essentiality. EMBO Mol Med 2015;7(4):368–79.
[91] Tolaney SM, et al. Phase II and biomarker study of cabozantinib in metastatic triple-negative breast cancer patients. Oncologist 2017;22(1):25–32.
[92] Tolaney SM, et al. Role of vascular density and normalization in response to neoadjuvant bevacizumab and chemotherapy in breast cancer patients. Proc Natl Acad Sci U S A 2015;112(46):14325–30.
[93] Mehta S, et al. Radiogenomics monitoring in breast cancer identifies metabolism and immune checkpoints as early actionable mechanisms of resistance to anti-angiogenic treatment. EBioMedicine 2016;10:109–16.
[94] Brodsky AN, Odenwelder DC, Harcum SW. High extracellular lactate causes reductive carboxylation in breast tissue cell lines grown under normoxic conditions. PLoS One 2019;14(6):e0213419.
[95] De Miguel MP, Alcaina Y, de la Maza DS, Lopez-Iglesias P. Cell metabolism under microenvironmental low oxygen tension levels in stemness, proliferation and pluripotency. Curr Mol Med 2015;15(4):343–59.
[96] Xiang L, Semenza GL. Hypoxia-inducible factors promote breast cancer stem cell specification and maintenance in response to hypoxia or cytotoxic chemotherapy. Adv Cancer Res 2019;141:175–212.
[97] Schito L, Semenza GL. Hypoxia-inducible factors: master regulators of cancer progression. Trends Cancer 2016;2(12):758–70.
[98] Ullmann P, Nurmik M, Begaj R, Haan S, Letellier E. Hypoxia- and MicroRNA-induced metabolic reprogramming of tumor-initiating cells. Cells 2019;8(6).
[99] Ivanovic Z. Hypoxia or in situ normoxia: the stem cell paradigm. J Cell Physiol 2009;219(2):271–5.
[100] Lan J, et al. Hypoxia-inducible factor 1-dependent expression of adenosine receptor 2B promotes breast cancer stem cell enrichment. Proc Natl Acad Sci U S A 2018;115(41):E9640–e9648.
[101] Yan Y, et al. HIF-2alpha promotes conversion to a stem cell phenotype and induces chemoresistance in breast cancer cells by activating Wnt and Notch pathways. J Exp Clin Cancer Res 2018;37(1):256.
[102] Abdollahi H, et al. The role of hypoxia in stem cell differentiation and therapeutics. J Surg Res 2011;165(1):112–17.

[103] Mohyeldin A, Garzon-Muvdi T, Quinones-Hinojosa A. Oxygen in stem cell biology: a critical component of the stem cell niche. Cell Stem Cell 2010;7(2):150–61.
[104] Huang X, Trinh T, Aljoufi A, Broxmeyer HE. Hypoxia signaling pathway in stem cell regulation: good and evil. Curr Stem Cell Rep 2018;4(2):149–57.
[105] Dello Sbarba P, Cheloni G. Tissue "Hypoxia" and the maintenance of leukemia stem cells. Adv Exp Med Biol 2019;1143:129–45.
[106] Ciccone V, et al. Stemness marker ALDH1A1 promotes tumor angiogenesis via retinoic acid/HIF-1alpha/VEGF signalling in MCF-7 breast cancer cells. J Exp Clin Cancer Res 2018;37(1):311.
[107] Baran N, Konopleva M. Molecular pathways: hypoxia-activated prodrugs in cancer therapy. Clin Cancer Res 2017;23(10):2382–90.
[108] Hunter FW, Wouters BG, Wilson WR. Hypoxia-activated prodrugs: paths forward in the era of personalised medicine. Br J Cancer 2016;114(10):1071–7.
[109] Spiegelberg L, et al. Hypoxia-activated prodrugs and (lack of) clinical progress: the need for hypoxia-based biomarker patient selection in phase III clinical trials. Clin Transl Radiat Oncol 2019;15:62–9.
[110] Hunter FW, et al. Identification of P450 oxidoreductase as a major determinant of sensitivity to hypoxia-activated prodrugs. Cancer Res 2015;75(19):4211–23.
[111] Kannan P, et al. Functional parameters derived from magnetic resonance imaging reflect vascular morphology in preclinical tumors and in human liver metastases. Clin Cancer Res 2018;24(19):4694–704.
[112] O'Connor JP, et al. Imaging biomarker roadmap for cancer studies. Nat Rev Clin Oncol 2017;14(3):169–86.
[113] O'Connor LJ, et al. Design, synthesis and evaluation of molecularly targeted hypoxia-activated prodrugs. Nat Protoc 2016;11(4):781–94.
[114] Dickson BD, Wong WW, Wilson WR, Hay MP. Studies towards hypoxia-activated prodrugs of PARP inhibitors. Molecules 2019;24(8).
[115] Mistry IN, Thomas M, Calder EDD, Conway SJ, Hammond EM. Clinical advances of hypoxia-activated prodrugs in combination with radiation therapy. Int J Radiat Oncol Biol Phys 2017;98(5):1183–96.
[116] Chen IJ, et al. Selective antibody activation through protease-activated pro-antibodies that mask binding sites with inhibitory domains. Sci Rep 2017;7(1):11587.
[117] Banerjee D, Harfouche R, Sengupta S. Nanotechnology-mediated targeting of tumor angiogenesis. Vasc Cell 2011;3(1):3.
[118] Khan KA, Kerbel RS. Improving immunotherapy outcomes with anti-angiogenic treatments and vice versa. Nat Rev Clin Oncol 2018;15(5):310–24.
[119] Chae YK, et al. Overexpression of adhesion molecules and barrier molecules is associated with differential infiltration of immune cells in non-small cell lung cancer. Sci Rep 2018;8(1):1023.
[120] Tan LY, et al. Control of immune cell entry through the tumour vasculature: a missing link in optimising melanoma immunotherapy? Clin Transl Immunol 2017;6(3): e134.
[121] Butters O, Young K, Cunningham D, Chau I, Starling N. Targeting vascular endothelial growth factor in oesophagogastric cancer: a review of progress to date and immunotherapy combination strategies. Front Oncol 2019;9:618.
[122] Yi M, et al. Synergistic effect of immune checkpoint blockade and anti-angiogenesis in cancer treatment. Mol Cancer 2019;18(1):60.
[123] Tian L, et al. Mutual regulation of tumour vessel normalization and immunostimulatory reprogramming. Nature 2017;544(7649):250–4.
[124] De Palma M, Jain RK. CD4(+) T cell activation and vascular normalization: two sides of the same coin? Immunity 2017;46(5):773–5.

[125] Shigeta K, et al. Dual PD-1 and VEGFR-2 blockade promotes vascular normalization and enhances anti-tumor immune responses in HCC. Hepatology. doi: 10.1002/hep.30889 2019.
[126] Allen E, et al. Combined antiangiogenic and anti-PD-L1 therapy stimulates tumor immunity through HEV formation. Sci Transl Med 2017;9(385).
[127] Zheng X, et al. Increased vessel perfusion predicts the efficacy of immune checkpoint blockade. J Clin Invest 2018;128(5):2104–15.
[128] Brooks JM, et al. Development and validation of a combined hypoxia and immune prognostic classifier for head and neck cancer. Clin Cancer Res 2019. Available from: https://doi.org/10.1158/1078-0432.CCR-18-3314.
[129] Fukumura D, Kloepper J, Amoozgar Z, Duda DG, Jain RK. Enhancing cancer immunotherapy using antiangiogenics: opportunities and challenges. Nat Rev Clin Oncol 2018;15(5):325–40.
[130] Datta M, Coussens LM, Nishikawa H, Hodi FS, Jain RK. Reprogramming the tumor microenvironment to improve immunotherapy: emerging strategies and combination therapies. Am Soc Clin Oncol Educ Book 2019;39:165–74.

Index

Note: Page numbers followed by "*f*" and "*t*" refer to figures and tables, respectively

B

C

W

X

Y

Z

Kolibri 360 GmbH
Ferdinand-Jühlke-Straße 7
99095 Erfurt, Deutschland
produktsicherheit@kolibri360.de